Storm Over Biology

Storm Over Biology

Essays on Science, Sentiment, and Public Policy

Bernard D. Davis

Adele Lehman Professor
of Bacterial Physiology, Emeritus
Harvard Medical School

Prometheus Books

Buffalo, New York

Published 1986 by Prometheus Books
700 East Amherst Street, Buffalo, New York 14215

Library of Congress Card Number: 85-63388
ISBN 0-87975-324-2

Printed in the United States of America

To my wife, and to a multitude of
friends, torn by tensions between the
heart and the head

Contents

Part III. Genetics, Racism, and Affirmative Action

Part IV. Medical Education and Affirmative Action

Part V. Public Concern Over Science

Part VI. Genetic Engineering

Foreword

Scientists and the Antiscience Movement

Edward Shils
(University of Chicago)

I

The antiscience movement that has grown up in recent years is ignored by most of the scientific community. But it derives much of its force from the support of a small group of scientists, and it may accomplish real mischief. That is what much of this book is about.

The disaffected scientists charge that the application of scientific knowledge is often pernicious, and some of its branches are even inherently dangerous. Sometimes coupled with this criticism is the view that modern science is driven by a desire to dominate both man and nature, and that it therefore seeks alliance with the earthly powers in the polity and the economy. Another frequent argument is that those who practice scientific research are members and supporters of "the elite," disregarding the welfare of the populace, excluding them from deliberations and decisions, and choosing problems without concern for the needs of ordinary persons. A more specific accusation is that certain branches of science believe in the inequality of human beings and attempt to demonstrate it.

Some critics further assert that scientific knowledge is in fact only a part of the ideology of bourgeois society. Being affected with interests—e.g., the pecuniary interest of capitalistic enterprise, the interest of the politician in acquiring and maintaining power, and science's own interest in domination—its pretensions to objectivity are alleged to be baseless. The detachment that has long been a source of pride for scientists would then be simply a pleasing fiction. An important school of the sociological study of science takes the position of Karl Mannheim—one of the fathers of the "sociology of knowledge"—that practical interest and "social position" enter into the very categories and the criteria of validity of knowledge.

Most scientists are probably not much interested in these metascientific or extra-scientific questions: they are quite satisfied and preoccupied by the intellectual challenges and practical responsibilities of their own research. They do interest themselves in the adequacy of the government's support of research and training; but scientists are not especially partisan in their complaints, nor do they espouse these negative assessments in any ideological mode. A good many scientists may also be attentive to the issues of "arms control," pollution of the atmosphere, and damage to the environment; but they do not place blame for these developments on scientists like themselves, nor do they think that such problems are inherent in scientific activity. Very few scientists are antivivisectionists, or think that the deceptive experimental manipulation of human subjects is a widespread practice; and few are alarmed by recent instances of fraudulent behavior in research. A relatively small number of scientists are interested in the philosophy of science in its newer wrinkles, and the question of whether objectivity is possible does not preoccupy them: the practice of research compels acceptance of objectivity as a postulate.

II

Nevertheless, there is a current of "antiscience"; its existence cannot be denied. Its proponents are both energetic and forceful. And, paradoxically, a small number of practicing scientists immerse themselves in it. But its primary adherents are drawn from the publicists of science—a sub-profession of journalism whose writers appear in the news sections of scientific journals, in daily newspapers, and in general periodicals. To be sure, this group expresses only specific beliefs of antiscience, in writing about particular events and policies; only a handful of those who swim in the currents of antiscience do so equally in all of them. Still, these various currents do represent a coherent body of beliefs.

Most scientists are not interested in the antiscience movement, and as far as they can see they are not affected by it. But this does not mean that antiscience is of no consequene, or that scientists, and the laity who appreciate the intellectual and the practical value of science, should be indifferent to it. We can by no means be certain that the beliefs contained in this movement are entirely without effect, or will be in the future.

One possible impact is on the support of science. Research is no longer the work of amateurs who pay for it out of their own pockets. Today it must depend heavily on the grant of funds by governments, who depend in turn on the willingness of the electorate to be taxed for this purpose. Thus far the public at large appreciates scientific research and its applications in technology, and so do most of its elected representatives. But should the electorate begin to disapprove, its representatives might become less ready to vote for extensive support. Of course, the electorate's disapproval, like its approval, is rather amorphous, and is slow and blurred in its effects. Nonetheless, the possibility of a shift in attitude exists and should not be disregarded. The antiscience movement could, if it continues and gains strength, contribute to turning the present amorphous approval around. It need not cause a stampede in order to have an injurious effect; small shifts can be disproportionately effective.

Another possible impact is on recruitment. The advance of science depends on a continuing flow of new entrants, who have to pass from the zone of society that surrounds the scientific community into the terrain in which that community is active. We are not guaranteed that there will always be a large number of zealous and talented aspirants for a life devoted to the cultivation of science. Should the beliefs of "antiscience" find increasing acceptance in the public, they are bound to affect in various and vague ways the flow of young men and women into careers in science. Of course, even if the wider public were to become less appreciative of scientific knowledge and its prowess in benefiting our lives, some individuals of strong character would still remain dedicated to this activity. But the pool would become smaller. This too would be damaging to the growth of scientific knowledge and its positive applications.

Finally, there is another important reason to attend carefully to the antiscience movement, and to sift the merit of some of its arguments from the dross in the rest. In the past two centuries scientists in Western civilization have been borne on a stream of growing public appreciation of science, in both its intellectual and its practical aspects. This appreciation has been a source of encouragement for both aspirants and practicing scientists, strengthening confidence that their undertaking is worthwhile. But scientists, despite the discipline that is integral to their profession, are also susceptible to the force of opinion; their morale is not indestructible.

We may see a warning in the erosion of the religious convictions and the morale of the Christian clergy in the present century, in the face of the steadily encroaching tides of secularization. The profession of science is not immune to a similar erosion of conviction and morale. Should public opinion begin to move toward the kinds of beliefs that are held as articles of faith by spokesmen for the antiscience movement, those beliefs would very likely seep further into the scientific community. As a result, more and more scientists might increasingly come to doubt the value of what they are doing. Such a development would inevitably do considerable damage to the growth of scientific knowledge.

III

The indifference of most scientists to the agitation mounted by the proponents of antiscience is understandable. In one aspect, it may be regarded as evidence that the morale of these scientists is still very strong. But that may not be the only reason for their silence. Their reluctance to express themselves publicly against antiscience may well be an indication that they share, if only with a part of their minds, some of those beliefs. Many American scientists are progressive in their social outlook, and antiscience often speaks with the rhetoric of progressivism. Like the Democrats described by Mr. Dooley in a certain ward in Chicago, who were so devoted to their party that they would sooner die than be buried by a Republican undertaker, many scientists cannot bear to dissociate themselves from the spokesmen for allegedly progressive causes. Though still strongly committed to the tradition of science, they do not wish to appear to be unsympathetic toward movements whose aims are clearly virtuous.

All the more admirable, therefore, has been the unceasing activity of Professor Bernard Davis in his affirmation of the best traditions of science. Professor Davis is no uncritical devotee of a narrowminded, hard-bitten scientism. He is neither a naive rationalist who thinks that scientific knowledge is the only legitimate kind of knowledge for human beings to pursue, nor a believer that this knowledge can supply exhaustive answers to questions of moral values. Moreover, Professor Davis is well aware of what is socially meritorious in the arguments of the antihereditarians, and he disapproves, with them, of discrimination on grounds of ethnic origin, religious belief, or social class. Yet, he does not believe that one can help overcome such discrimination by suppressing research into the genetic diversity of human beings, or by relaxing the standards for assessing academic and scientific achievements. He also recognizes that the progress and the applications of scientific knowledge have given rise to problems that are as yet unsolved, and he is not oblivious to the moral sensibility underlying the criticisms made by the representatives of antiscience. He will have none of their derogation of the value of objective scientific knowledge.

Professor Davis has been criticized and even censured for his brave pronouncements, but he has not been deterred from returning again and again to reaffirm the value of scientific objectivity, the freedom of scientific research, and the practical application of scientific knowledge as far as it can be carried within the limits of reason and humanity. He has done so with courtesy and consideration for his adversaries, and for colleagues who agreed with him but were too timorous to come to his defense. Moreover, he has set forth his arguments with a lucidity and refinement of expression that has become as rare as courage. For all of these virtues he deserves our commendation.

The large number of essays that make up this book present straightforward analyses of many of the complex and difficult problems that beset scientific research and teaching nowadays. I hope that his example will be taken to heart, and that his views will contribute to a deeper understanding, among scientists and nonscientists, of the value of scientific knowledge and of the power and responsibilities of those who create it.

Preface

Recognizing that science and technology have become major outlets for the expression of the creative spirit, and that they are also ultimately responsible for the extraordinary increase in the rate of cultural change in the modern world, laymen are more and more interested in their advances. Excellent descriptions are therefore now available in many publications—for example, *Discover, Omni, Science '86,* and the science pages of leading newspapers.

However, it is even more important for the responsible citizen to understand the nature of science, as a methodology and as an enterprise, since its interactions with society press on us increasingly. Examples include decisions about the support and the regulation of science, about evaluating its possible dangers, and about determining its place in general education and in the formulation of public policy. Here it is more difficult to obtain a reliable basis for judgment. Morever, we have been late in recognizing that science, and technology even more, create costs and dangers as well as benefits. As a result, participants in the new academic discipline of science and society, as well as science writers who publish in popular science magazines, have tended to concentrate on this aspect of the subject, giving the public an unbalanced picture.

This picture has been further distorted by a curious development in the study of the history of science, which spread from the Soviet Union in the 1930s: the claim that because socioeconomic forces influence the course of science (which is obviously true), then the alleged objectivity of science—which has been traditionally considered the heart of the enterprise—is a myth. This theme has appealed not only to Marxists, who find it a help in dismissing contradictions between theory and practice, but also to many historians of science in the West. Bored with recording how the solution of a scientific problem generates its successors ("internalism"), they have developed an "externalist" school whose excessive emphasis on social factors has made it fashionable to question the objectivity of science. From another direction, many social scientists have further reinforced this skepticism: frustrated by their difficulty in achieving objectivity, they have attempted to strengthen the similarity of their field to the natural sciences by denying that either could be truly objective. Meanwhile, working natural scientists continue to produce results that clearly reflect reality, and by and large they ignore these criticisms of objectivity that pour in from the periphery of their field.

Concerned that much of the literature on the interactions of science and society rarely reflects science as scientists see it, I have published on a number of topics in this area, and in a variety of places, over the past fifteen years. These papers have now been collected in the present volume. The thread that connects them is the issue of objectivity, and so I begin with a paper on this topic. Most of the other papers are less technical and focus more on social problems.

In this volume I frequently consider the limits of science—limits that we encounter as soon as we ask questions that involve values. These questions preoccupy most people far more than questions about the nature of the external world, and with good reason. But even though science cannot prescribe answers to our many questions of values, it can play a useful adjuvant role and contribute toward finding better answers. Recent publications in the newly developed field of sociobiology, branching from evolutionary biology, have gone farther and suggested that biology may be able eventually to make more direct contributions to ethics. Several of the essays in this collection discuss sociobiology, defending it against political attacks, but at the same time offering criticisms that are rather different from those that others have voiced.

Many of the papers included here relate to other aspects of evolutionary biology—the field that unifies all of biology. In one essay I argue that we could make the teaching of evolution more convincing, and more interesting, if we built more on the recent, direct evidence from molecular genetics. For example, without the evolutionary concept of the genetic continuity of the entire living world we would not be able to use the molecular biology of bacteria to help us understand human cells, nor would we have a bioengineering industry based on the transfer of genes between distant species.

Since evolution depends on selection from a reservoir of variation, it carries the important implication that our species possesses wide genetic diversity in all the traits that have changed in the recent stages in our evolution—including various abilities and other behavioral traits. Of course, because this topic is directly connected to public policy it has inevitably aroused strong feelings—but that connection is also why it is important. Ironically, the source of the opposition is secular rather than religious, and it lies largely within the academic community. Moreover, its egalitarian arguments against recognizing genetic limits appeal to our warmest and noblest sentiments. Hence, even though the result of this opposition has been virtual repression of the field of human behavioral genetics, the attacks have gone largely unanswered by biologists. Because of this vacuum I have given much more attention to this topic than to the problem of creationism.

Another set of papers deals with quite a different aspect of genetics: our growing power to manipulate genes. Here I try to assess the probability of various projected future developments (especially possible applications to humans). I also consider the thorny question of the roles of the scientific community, the public, the media, and the courts in trying to reach sensible decisions on such highly technical issues.

Finally, one group of papers concerns the recent widespread retreat of our society from another kind of objectivity—simple honesty in evaluating individual performance—as a consequence of the transformation of affirmative action into reverse discrimination. Unlike most widely shared flights from reality, which have arisen from commitments to either theological or ideological dogma, this one arose from widespread guilt. This guilt was thoroughly justified, and it was necessary to move us to eradicate our racist legacy of slavery; but it also led, as any powerful emotion can do, to some irrational and unrealistic positions. The intense feelings on this subject led to quite a storm when I published a criticism of medical schools for sacrificing standards beyond reason in order to fill minority quotas. I have now decided to present the history of this episode, for reasons that I spell out. Unlike the rest of the volume, this piece (number 23) has not been previously published.

In closing, I would note that the recent court decisions about creationism and teaching clearly reflect an increased public acceptance of the vital connection between science and the purposes of education in a democratic society. This development encourages confidence that science will also ultimately prevail in the other, more challenging controversies addressed in this book, in various areas in which science intersects with public policy. As I have noted repeatedly, nature will have the last word.

This collection ranges from short editorials, book reviews, and letters to substantial essays. In this wide range of topics the reader will readily discern a unifying theme: dedication to the objectivity of science in deepening our insights into reality, and to the value of incorporating the resulting insights in the formation of public policies. Though I use the term "essays" in the subtitle, these papers are not essays in the sense of a ruminative, entertaining, and urbane literary form. They are expository and analytical statements, not enriched by vivid imagery but written in the style recommended by Francis Bacon: "Science must be writ plain." But, of course, this approach has a certain dryness. As Bacon also wrote, poetry "doth raise and erect the mind, by submitting the shews of things to the desires of the mind; whereas reason doth buckle and bow the mind to the nature of things."

I have made a few stylistic changes in some of the original versions and have also eliminated gross duplications. In addition, I have added an introductory comment before each paper to provide continuity and historical perspective, and occasionally to update my position. To help the reader choose between shorter and longer treatments of the same topic these comments include, where appropriate, brief synopses.

Responses to these published writings have sometimes made me feel that I was in a position aptly described by E. B. White as membership in a party of one. It has been painful to offend friends whose positions on many other social issues I share. I can only hope that this volume will encourage a search for ways to combine humanitarian ideals with a tough-minded recognition of reality—a position that cannot be accurately mapped on a linear scale of left to right.

I am grateful for helpful comments from members of my family, and from David Heilbroner and Margaret Olmsted. I benefited greatly from the hospitality of the Center for Advanced Studies in Behavioral Sciences at Stanford University in 1974-75, where I wrote some of these pieces and acquired the background for others.

Part One

Objectivity and Science

1

Science, Objectivity, and Moral Values

Underlying all the later pieces in this volume is the conviction that our social policies will be most effective if we base them on objective knowledge of reality. This paper addresses itself to the question of whether science really gives us such knowledge.

After its publication, I found that Robert Merton had presented closely related ideas (in a chapter entitled "The Normative Structure of Science," from his book The Sociology of Science, *University of Chicago Press, 1973; originally published in the* Journal of Legal and Political Sociology *1 [1942]: 115). That essay is justly famous for analyzing the ethos of science in terms of four institutional imperatives: universalism, "communism," disinterestedness, and organized skepticism. More relevant for the present essay is the further comment that "science is a deceptively inclusive word," commonly used to denote several different things: a set of characteristic methods, a stock of accumulated knowledge, a set of cultural values and mores governing the activities termed scientific, or any combination of these.*

In the context of evaluating the objectivity of science, however, this classification of the several meanings of the term does not sufficiently highlight the contrast between its subjectivity in one meaning and its objectivity in others. For while the practice of science regularly involves highly subjective value judgments, both by the individual scientist and by the supporting institutions, the main thesis I develop here is that the presence of such subjective features in the activity does not weaken the objectivity of the product. While this point seems obvious, failure to recognize it has clearly been the source of much confusion in philosophical and sociological discussions as to whether science is or can be value-free.

Modified from *Program on Public Conceptions of Science*, Harvard University, Newsletter, April 11, 1975; now *Science, Technology, and Public Policy*, MIT Press.

I also take up a related topic, namely, the implications of science for morality. While the formulation of rules of morality, as well as individual decisions, clearly cannot be entirely objective, I point out ways in which the methodology of science is nevertheless relevant here, even though not prescriptive.

Some years ago C. P. Snow[1] and Jacob Bronowski[2] tried to close the gap between the two cultures—science and the humanities—by focusing on their shared esthetic values, and especially on the similarities between scientific and artistic creativity. Increasingly, however, the world has shifted attention to the relations of science to moral rather than to esthetic values.[3] These relations have turned out to be highly controversial.

A few decades ago it was widely believed that the remarkable success of science in dealing with increasingly complex questions, from Newton's laws of motion to the nature of the gene, could be extended without limit: the powerful tools of scientific methodology, applied to social studies, should eventually (and perhaps even soon) provide correct solutions to the major problems of society. Today, however, this assumption (sometimes called scientism) is obsolete. As Peter Medawar[4] in particular has pointed out, science can solve only problems about the nature of the external world—problems for which there exist, in principle, correct answers. In contrast, problems involving moral or esthetic values have no objectively correct solutions, except in the sense of historical accuracy or of conformity to legal or other social conventions.

But if we agree that the scientific method can solve only certain kinds of problems, we are nevertheless not so clear about where the boundaries actually lie, and whether science can help us to build social policies more firmly on reality. One group of critics—including neo-Marxists, historians of science who overemphasize the "externalist" interpretation of the nature of scientific discovery, and nihilists of the counterculture[5]—consider science so dominated by the political and economic values of the surrounding community that its alleged objectivity is a myth. A second group[6] would accept its fundamental objectivity but would deny or minimize its relevance to social problems, because they view these as challenges essentially to our moral judgment. I wish to examine the validity and the implications of these views.

Objectivity in Science

To clarify the question of the objectivity of science it is essential to recognize that the word *science* is used with three different meanings, depending on the context. First, science is a *methodology* that aims at achieving maximal objectivity, in two general, interacting ways: by relying only on verifiable observations, logical inferences, and predictions that are tested against the external world and not against our social values and preconceptions; and by employing extensive communication, in a disciplined and informed community, to check the observations and to critically assess the inferences. The second meaning of *science* is a coherent, growing *body of public knowledge*, resulting from the cumulative application of the methodology. These two are the usual meanings in discussions of the philosophy of science.

In a third quite different use, the term *science* refers to a set of human *activities*, in what has been called the context of discovery rather than the context of justification. In contrast to the first two aspects of science this one is indeed heavily value-laden: in a paradox pointed out by Beveridge,[7] the practice of science is an art. Thus the individual investigator, in deciding what to study and how to study it, is constantly making value judgments. Similarly, when he constructs concepts and hypotheses he is engaging in a subjective act of imagination. Nonobjective features are also found in much of the sociology of science: the scientific community incorporates esthetic and pragmatic values in judging the importance of various discoveries, and social institutions do likewise in determining how much support to provide for various areas of research. Clearly, science as an activity contains major nonobjective elements essential for its creative function. But their presence does not undermine the ability of science to yield an objective body of knowledge.

One could add the caveat that the whole scientific methodology rests on tacit assumptions about the existence of a real, material universe, and also about experience as a source of knowledge. But these are *a priori* axioms, residing at a different metaphysical level, and their implications for the objectivity of science would be outside the scope of this paper.

The importance of corrective feedback in promoting objectivity in science cannot be overemphasized. The phenomena studied are complex, observers and instruments are fallible, and human beings are inevitably tempted to prefer findings that support their preconceptions. Hence we must emphasize that at the growing points of science its objectivity is often imperfect: indeed, and many published observations and conclusions fail to hold up and are later discarded by the scientific community (usually by ignoring rather than by publicly correcting them). But such revelations of error do not mean that science as a whole is arbitrary and subjective: the process of critically assessing the important findings by the scientific community is remarkably efficient. Moreover, in contrast to absolutist systems of belief, the scientist's version of truth, even when widely accepted, occasionally has to be modified in the light of further advances.[8]

In a word, though the regions of growth in science are fragile and often beset by conflict and contradictions, their continual crumbling and repair are not indices of weakness in the foundations. Were they shaky, so elaborate an edifice would collapse. On the contrary, the coherence of the growing body of science, and the success of its predictions and its technological applications, continually affirm the objective validity of the earlier knowledge that is being built upon.

According to this view, then, the preferences and prejudices of society should have very little influence on the *reliability* of what the scientific community normally accepts as true in the natural sciences. (Exceptions, however, will be considered in the next paragraph.) To be sure, social forces do influence the pattern of growth of scientific knowledge, especially through control over financial support; but even this influence affects primarily the rate, and not the direction, of advance. As Steven Weinberg has noted,[9] science progresses chiefly by asking those questions that are most fruitful at a given stage in its development, and such questions are largely generated by preceding discoveries rather than by social pressures. An interesting example is the recent crash program in cancer research, which was seen by many as a type of social tumor, generated by political pressures and released from the normal controls of the scientific community over the direction of growth. Nevertheless, the content of the program stemmed from the recent emergence of several new fields (molecular genetics, virology, and cell biology): it could not have been envisaged twenty years earlier.[10]

Social pressures can thus influence only the *selection* of the scientific truths that are discovered, and not their validity. However, these pressures can also create biases, and strong resistance to the acceptance of valid conclusions, especially when a branch of science is thought to conflict with a theological, moral, or political doctrine—witness Galileo, or the fate of Soviet genetics under Lysenko. But again, in contrast to conflicts in many other areas of human endeavor, scientific questions can finally yield correct solutions, determined by what is out there in nature—a solution quite independent of the political and cultural pattern of the country in which it is discovered.

To be sure, our measurements of what is out there are often imprecise, especially when the phenomena are complex and the tools crude. But we should not confuse imprecision with lack of objectivity. All scientific measurements have characteristic limits of precision, which vary widely from one kind of measurement to another; and demands for perfectionists, unattainable standards, inappropriate for a particular field, are usually a sophisticated means of expressing resistance based on emotional attitudes. Lest it be thought that we have outgrown such problems, I suggest that the reader consider whether these statements apply to the current conflicts over genetics and intelligence.

It is thus clear that scientific truth is fundamentally different, in its emphasis on a consistent objectivity, from political or religious or artistic truths, in which questions of meaning and value play a predominant role. In fact, *by definition* no source of objective knowledge about the nature of the universe can be distinct from science: any novel approach that adds to such knowledge, and any question that moves from speculation to an objective answer, automatically becomes part of science. For example, the problem of the nature of space and time has moved from metaphysics to relativistic physics. Similarly, until this century, our ideas about the nature of races, assigning to each a distinct and uniform set of characteristics, were based on the Platonic concept of ideal types, defining each category of entities. Modern population genetics, however, has now replaced this view. We now recognize that every natural population includes a wide range of genetic patterns, and so its structure must be described and understood in statistical rather than stereotypic terms.[11]

While recognizing that science derives great power from its objectivity, we must equally recognize that it has severe limits. In fact, there is surely much truth to the accusation that the glaring success of science has deepened the shadows surrounding other intellectual activities, including ones that are more important in our daily lives: those concerned with profound human needs and with the search for kinds of meaning and truth that nourish the spirit (if I may use this term to sum up a set of emotional and social patterns far too complex for description in scientific terms). Perhaps we will be able to achieve a better balance between the two activities, and reduce tensions, if we can define more clearly, and can accept, those regions that are appropriate for science, those appropriate for the humanities, and also those where the two approaches can complement each other.

Relevance for Human Affairs

This challenge brings us to the second criticism of science: that its method and its results, however objective, have little relevance for moral and political decisions. Indeed, since these decisions do involve value judgments, science by itself cannot specify the answers, as we have already noted. However, our decisions depend not only on value judgments, but also on estimates, both of the methods for reaching alternative goals and of the consequences; and the scientific method is especially good at the empirical predictions involved in such estimates. Hence, while science cannot *prescribe* correct answers to such questions, it is nevertheless *relevant*.

This ability of science to complement morality and political theory as guides to action should not be minimized. For the increasingly rapid changes in technology and social patterns inevitably diminish confidence in our ability to predict the long-term consequences of our actions; and this loss of confidence surely has contributed to the anxiety of our age. We should therefore not cast aside lightly tools that can improve our predictions, even though they cannot provide complete answers.

Moreover, we make our individual decisions within a framework of social values; and just as science is relevant to our decisions, it also can influence, the process of developing the underlying framework, in several ways. First, the methodology of science places a premium on intellectual independence and on objective truth, thereby generating a skeptical attitude toward authority. The spectacular success of this methodology has inevitably caused its values—what Bronowski has called "the habit of truth"—to spill over to some degree to the rest of our culture. Second, the methodology of science also promotes democracy, since its attention to the quality of a person's evidence and reasoning has decreased the traditional respect for hierarchical authority and for inherited status. Third, in some areas of political decision-making the rise of science has decreased the role of those with ideological qualifications and increased the role of those with specific expertise.[12] Finally, while intellectuals have tended in recent years to focus on the previously hidden costs of modern technology, and to take for granted its contributions to economic advances and to the spread of education, it is obvious that these contributions have also been the prime sources of the modern sense of broadened obligations: to eliminate poverty, to distribute opportunities more fairly, and to develop a more egalitarian social climate.

Evolution and Ethics

A specific consequence of the scientific method, the discovery of man's biological origin by natural selection, has had even more direct implications for our ethical systems. Unfortunately, these implications were distorted by the early Social Darwinists, who focused purely on the competitive aspects of our biological legacy and used this as an analogy to rationalize prevailing economic practices. Today, however, the emerging science of sociobiology[13] is emphasizing the importance of social cooperation as well as competition in evolution, and it seems certain that future students of ethics, and of behavior in general, will be paying increasing attention to our biological roots. I shall here consider briefly several of their implications.

First, for those who prefer a world view that is thoroughly consistent with scientific knowledge of reality, Darwinian evolution has made it impossible to retain the long-cherished belief in a transcendental, supernatural basis for ethics.[14] Instead, in this view, when our species evolved a nervous system capable of the abstract thought and the elaborate language that led to our rich cultural evolution, and to large, complex social units, ethical systems would then have developed as the distillates of a great deal of experience in trying to reconcile individual impulses with the survival and welfare of the group. This view also predicts that ethical principles, as a product of cultural evolution, will shift with changes in social patterns. Indeed, they clearly have done so: the rapid social changes generated by modern technology have challenged many ethical principles that long seemed immutable, especially in the area of sexual mores.

A second contribution of evolutionary biology, in the new branch called sociobiology, is to recognize the biological foundations on which our cultures have developed ethical systems. Here linguistics provides a helpful model. For though we are born with the capacity for language, rather than with a particular language, Noam Chomsky has shown that all languages share deep characteristics; and since these are universal in our species they must be determined by inborn structures in the human nervous system.[15] Similarly, though we are not born with structures in our brain that specify a particular ethical code, our evolutionary survival as social animals requires us to have the capacity and the drive to develop a code that will balance social and individual interests.

This insight provides an alternative to the view of those philosophers and religious leaders who still seek a transcendental source of ethics, because they fear that if knowledge of evolution eliminated this source the only logical alternative would be the bleak one of a purely relativistic ethics. On the contrary, the inborn social drives studied in sociobiology provide a scientific base for another option—what philosophers call an intuitionist basis of ethics, and Freud called a superego. This biological legacy of cooperative, *self-transcendent drives*, required by our biological evolution as social animals, can explain why the social patterns and ethical systems produced by cultural evolution, however varied, have retained major common features.

A third inference from evolution is that ethical systems arise as compromises between our cooperative drives and our individualist, competitive ones, just as physical traits generally evolve by selection for optimizing compromises rather than for maximization. Moreover, individuals differ in their proportions of these conflicting drives. From this view it would follow that the research of some philosophers and theologians for a rigorously consistent, eternal set of detailed ethical principles is a search for a kind of philosopher's stone. For if we take seriously our origin by Darwinian, undirected evolution, we must recognize the basis of ethics as fundamentally relativistic, but within limits set by our biology as a social species. Hence any permanent features of ethical systems can only be very general ones.

In this view, E. O. Wilson[13] is also seeking the impossible in expressing confidence that science can ultimately lead to a "true" ethical system, fitted to our biological needs. His conviction is based on the knowledge that emotional drives, which ethical systems seek to moderate, arise largely in the limbic system in the brain, and the advances of neurobiology will ultimately give us detailed knowledge of this system. But since different individuals have different limbic systems, I have difficulty with the idea of a "true" ethics that will fit everyone's limbic system: like science in general, as I noted above, neurobiology also will be relevant, but not prescriptive, in our process of formulating ethics by social negotiation.

Finally, I wish to comment briefly on another sense in which it has been claimed, in recent years, that science is not value-free. This claim asserts that scientists have a moral obligation to evaluate the beneficial and the harmful consequences of their discoveries, and to prevent harmful uses and even avoid dangerous discoveries. And certainly scientists, with their earlier contact with the problems that they generate, and with their special knowledge of its technical aspects, have an obligation to try to educate a wider public. But if they went farther and tried to play a prescriptive role in judging benefit and harm, they would properly be accused of arrogating a responsibility that belongs to the whole community.

Conclusions

In summary, I have emphasized that the highly subjective features of science as an activity do not necessarily diminish the objectivity of science as a methodology or as a body of knowledge. Moreover, the activity inevitably yields frequent deviations from objectivity in the knowledge at the growing points—but the methodology provides mechanisms for weeding these out.

I would further suggest that science and morality are neither closely coupled nor completely uncoupled. As David Hume pointed out, one cannot derive an "ought" from an "is." Nevertheless, our "oughts" are not created in a vacuum: they derive *in part* from what is, as well as from traditions and from the subjective reactions of socially interacting individuals. Hence science, by making our perception of reality more reliable, and by improving our ability to predict consequences, can contribute to developing both more effective ethical systems and more effective individual responses, even though it cannot specify correct ones.

On the other hand, this partial coupling is not symmetrical: one cannot derive an "is" to any extent from an "ought." When one tries to do so one is engaging in a corruption of science—vividly illustrated by Lysenkoism, but also seen in milder forms in our country today, especially in issues where genetics and evolution are believed to threaten religious or political convictions.

I would also suggest that as our world grows smaller, and modern means of communication and transportation now make us all neighbors, the interactions of science and morality as guides to social and political action are becoming increasingly important. While humanitarian impulses are invaluable in guiding relations between immediate neighbors, they are less adequate guides in framing our responses to such global problems as the uneven geographic distribution of technological, economic, and demographic resources. In responding to such horrors as localized mass starvation, for example, we do not wish to abandon values whose roots are deeply entwined with our concepts of humanity and of civilization. Yet if we approach such problems only on a hand-to-mouth basis and fail to use the tools of science and technology to seek realistic long-term projections and solutions, we will generate a novel kind of moral culpability—less personal, but also more serious if our actions cause the problem to grow.

Finally, I would suggest that broader public education in the scientific mode of thought, with its deep respect for objectivity, might be helpful in decreasing international tensions. Through its universality, this approach can reveal areas of agreement, and shared challenges, that may help overcome separations created by other conflicts of interest and traditions. Without such an anchor to a common reality, and with more emphasis on subjective values, we are more likely to drift apart. If we choose to think only with our hearts, however sincerely, we will be approaching our conflicts of interest on the basis of emotionally charged beliefs; and this can lead to what has been called thinking with our blood.

Notes and References

1. C.P. Snow, *The Two Cultures and the Scientific Revolution* (London: Cambridge University Press, 1959).
2. J. Bronowski, *Science and Human Values* (New York: Harper Torchbooks, 1956).
3. "Science and Its Public: The Changing Relationship," *Daedalus* (Summer 1974).
4. P. Medawar, *The Art of the Soluble* (London: Methuen, 1967).
5. E.g., T. Roszak, *Where the Wasteland Ends* (New York: Doubleday, 1973).
6. E.g., S. E. Luria and Z. Luria, *Daedalus* (Winter 1975): 273.

7. W. I. B. Beveridge, *The Art of Scientific Investigation* (New York: Norton, 1950).
8. T. Kuhn, in *The Structure of Scientific Revolutions*, has emphasized the key role of revolutions, replacing earlier "paradigms," in the advance of science. This thesis has had a great impact on historians of science. But while it fits some major discoveries in physics in the first decades of this century, I believe most scientists do not find that it fits the way their fields advance. In biology, in particular, most advances are clearly refinements, answering previously open questions rather than overthrowing earlier dogmas. For example, the most fruitful advance in biology in this century has clearly been Avery's discovery that DNA is the genetic material, followed by Watson and Crick's discovery of the structure that accounts for its remarkable properties. But these developments filled a black box, rather than having to overthrow accepted concepts.
9. S. Weinberg, *Daedalus* (Summer 1974): 33.
10. Added note: The subsequent history of cancer research, since the original publication of this paper, reinforces the point. The politically dramatic cancer crusade produced relatively little return, in proportion to its heavy funding, until the recombinant DNA technology appeared. Remarkable insights then emerged, including recognition that alterations in certain cellular or viral genes (oncogenes) can release the uncontrolled growth characteristic of cancer cells. But the recombinant technology did not emerge as a planned product of the cancer crusade: it came, quite unpredictably, from quite independent esoteric studies in bacterial genetics.
11. The resulting picture of wide genetic diversity in our species seems to many to threaten current drives for greater social equality. In time, however, it will have to be accepted as a reality, since we cannot legislate it out of existence. Moreover, we can learn to build on this biological understanding in positive ways: to eliminate ancient ethnic prejudices and their pseudoscientific basis, and to improve education by adapting its procedures to the different capacities of different individuals.
12. H. Brooks, *Daedalus* (Winter 1965): 66.
13. E. O. Wilson, *Sociobiology* (Cambridge, Mass.: Harvard University Press, 1975).
14. J. Monod, *Chance and Necessity* (New York: Alfred Knopf, 1971).
15. N. Chomsky, *Aspects of the Theory of Syntax* (Cambridge, Mass.: MIT Press, 1965).

2

Ezra Pound, the American Academy of Arts and Sciences, and Intellectual Freedom

This piece takes up a specialized philosophical issue: the definition of intellectual freedom. The issue arose when a committee of the American Academy of Arts and Sciences recommended that the academy's Emerson-Thoreau Medal, for a lifetime of service to letters, be awarded to Ezra Pound. The Council of the Academy, of which I was a member, ordinarily accepts the recommendations of its committees for such awards. But while the committee had deemed it appropriate to recommend the award to Pound on the basis of his literary skill, despite repugnance for some of his ideas, many members of the council felt that it had to consider the justification in broader terms. The debate was extensive and soul-searching, as we all wrestled with this deep conflict of values. The decision, by a small majority, was to reject the recommendation.

The actions of the council are supposed to be confidential. However, as was inevitable with such a large group and such a highly charged issue, the news leaked out, and it aroused much discussion in academic circles in Cambridge. One member of the academy publicly accused the council of assaulting intellectual freedom and resigned. Since that member is Jewish, and therefore presumably would be as sensitive as I to the immorality of Pound's antisemitism, I found it hard to understand his absolute separation of form from ideas, and of intellectual from moral evaluation, in determining the criteria for making an award. I therefore wrote this piece in an effort to clarify the issue and to settle rumors about what had gone on in the council.

Guest editorial, *The Boston Globe*, August 7, 1972.

Dr. Jerome Lettvin has accused the Council of the American Academy of Arts and Sciences of assaulting intellectual freedom in rejecting a committee's recommendation that the Emerson-Thoreau Medal be awarded to Ezra Pound. I was privileged to participate in the extensive, profound, and soul-searching debate of the council; and even though it would be kinder to Pound to avoid discussing the reasons for the rejection, the issues are so consequential that they must be more fully explored.

First of all we must try to be clear as to what the award of the medal would mean and what denying it means. It is ludicrous to consider this denial in any sense comparable with book burning, or with preventing Solzhenitsyn from publishing his work, or with confining Zhores Medvedev to a psychiatric institution. The decision did not deprive Pound of the right to have his works published or to have them admired; neither did it convict him of a crime or deprive him of his liberty. Indeed, since it was a private decision it was not intended even as a criticism that would reach Pound or the public.

We must also recognize that the issue facing the council was not whether to condemn Pound, or even whether to honor his poetry; its task was strictly to decide whether or not he was an appropriate recipient for the Emerson-Thoreau medal. Indeed, subsequent to the council's negative decision some interested individuals began to explore the possibility of creating a more suitable mechanism, such as a scholarly conference and publication, that could honor specific literary contributions of Pound. There is little doubt that the council would have approved such a proposal. Unfortunately, while these discussions were underway a public outcry was precipitated by Lettvin's acute crisis of conscience.

Why did the council reject the committee's recommendation? First, Pound's image struck some of us as incompatible with the humanitarian tradition of Emerson and Thoreau. A more fundamental consideration is that this medal honors an individual for his total contribution and is not a prize given for a specific literary work. It seems to say to the public that this is a great man, to be emulated. And though the award committee based their recommendation specifically on Pound's craftsmanship, creativity, and germinal influence as a poet, the council felt that any explanations accompanying the award, however elaborate, could not eliminate the implication of approving the recipient's life as a whole. Since Pound's antisemitic and profascist ideas are perhaps better known to the general public than is the beauty of his poetry, the award might be widely construed as saying that these ideas are not so bad after all.

This is the nub of the matter. Despite the increasing moral relativism of our times, the majority of the council took the position that there are moral standards worth upholding, and that some views must therefore be condemned as evil. We would defend the right of such views to be published, and to be debated; but we saw no need to honor them or even to give the impression that we considered them acceptable. Most members of the council are of a generation that was united by the struggle to prevent the Nazi horror from encompassing the world. Moreover, the evil of racism is still very much alive, and is one of the critical issues of our times—witness the problems of racial integration in our country today, and the racial aspects of the Vietnam war. Under these circumstances the possibility of appearing to condone racism seemed to some of us a very serious consideration.

Two further points seemed significant, though less fundamental. First, Pound's medium as an artist was words, which he used to express ideas as well as feelings and esthetic insights. Hence to consider form alone, and to ignore the content of his ideas, would be harder to justify than if one were considering an award, say, to a musician. Secondly, we must recognize the limitations of an academy—a body of people with shared interests but a variety of views. It is questionable whether the council of an academy has the right to make an award in its name that would surely be deeply offensive to many of its members.

The council thus faced an issue to which more than one moral principle applied, and these principles were in conflict. Members differed in their decision as to where wisdom lay. Some had strong convictions on one side or the other, some felt sorely torn, and some changed their minds in the course of the discussion. Since I am not writing officially I feel free to express my surprise at the cheapness with which some who should know better have felt free to apply a single, simplistic yardstick and to castigate the council for not having reached the "right" decision in so complex an issue.

3

The Moralistic Fallacy

A preceding editorial in the distinguished journal Nature *suggested that we should limit the study of human behavioral genetics because its results might be too threatening to our social values. When I wrote a letter criticizing that conclusion the journal suggested that I expand it into a guest editorial, which follows. I take up the same question at greater length in Part Three, but I place this piece here because it poses the problem in a philosophical framework and it proposes a novel philosophical term.*

The increased focus of our age on social justice, and on the need to control the costs of technology, has had admirable consequences. But it has also reactivated an old threat to science: the demand that certain kinds of scientific knowledge be forbidden. George Steiner, in his recent Bronowski Memorial Lecture, rejected this proposal on the pragmatic grounds that it just won't work. However, *Nature*'s editorial of 2 February [1978] drew a different conclusion: that we are now groping for a code to protect society from dangerous knowledge, much as we have developed ethical codes to protect the subjects of biomedical research.

Since this issue is of central importance for the future of science, we must consider the arguments very carefully. I wish to discuss some that did not appear in Steiner's lecture or in *Nature*'s editorial. I shall focus on the heritability of intelligence, which Steiner views as the most intractable among the several kinds of dangerous scientific knowledge.

First, the analogy to medical research proposed in the editorial, though plausible at first glance, in fact ignores a crucial distinction: be-

---

tween actions that are themselves dangerous and knowledge that might lead to dangerous actions. Medical investigators do indeed accept ethical limitations on dangerous procedures, that is, those that would expose an experimental subject to loss of dignity, pain, or risk. And investigators in behavioral genetics are subject to analogous limitations: they cannot mate humans at will, or transfer identical twins into different homes, even though these procedures would be powerful tools for advancing knowledge. On the other hand, medical investigators are not forbidden to seek knowledge simply because its prognostic implications may be painful, for the subject or for others. So medicine does not provide a model for justifying limitations on the knowledge sought in other areas.

The very concept of dangerous knowledge is also shaky. Ever since the discovery of fire, and of cutting tools, it has been clear that virtually any scientific knowledge can be used for good or for ill: the costs and benefits depend entirely on how it is used. Moreover, we have only a very limited ability to foresee the eventual scientific benefits of a new discovery: science is a continuous web, and fundamental advances often arise through unexpected cross-fertilization. For example, there are very good reasons to forbid human cloning: but if we should forbid any research in cell biology that might bring cloning nearer we would seriously impair advances in cancer research. We must therefore ask whether it is more rational to try to protect society by limiting the areas open to fundamental inquiry, or by focusing on earlier assessment and improved control of new technological applications of scientific knowledge.

We must also consider the rather ahistoric and absolutist conception of justice implied in the suggestion of a fundamental incompatibility between man's hopes of justice and decency and certain categories of truth. For though it is clear that the concept of justice has certain stable features, it is also clear that the rules of behavior in any society, and the assumption underlying these rules, are continually evolving—especially when the society is faced with new knowledge and new technologies. For example, we have weathered the storm created by Darwin; and though the supernatural basis for a moral consensus was shattered by his elimination of special creation, we have meanwhile developed a radical increase in our sensitivity to problems of human rights. Can we not trust posterity also to adapt its notions of morality to further new knowledge?

More specifically, there seems to be a pervasive fear of the social impact of genetics, arising largely from the pseudoscientific extrapolations of Social Darwinism and Nazi racism. I would suggest that this view does not reflect the real contributions of this field. In fact, one of the historical grounds for racism was the prescientific conception of races as permanent, distinct products of creation. But evolution made us aware of the brotherhood of all races. Another rationale for racism was the typological conception of the nature of race—a view based on the Platonic characterization of groups in terms of an ideal "type," subject to only minor deviations in concrete individuals. But this misconception was destroyed by population genetics, which demonstrated the great genetic heterogeneity of all races, the statistical nature of the differences between them, and the extensive overlap of these distributions for most traits. This field has thus contributed far more than is recognized to public awareness that one cannot determine an individual's capacities by identifying him with an ethnic group.

Finally, we must ask how much the current reasons for proscribing an area of knowledge really differ from those used by Pope Urban VIII, Bishop Wilberforce, or Lysenko. The main difference is that science is penetrating increasingly into areas that generate moral problems, and that generate technological capacities for great destruction. Since Steiner considered the former the more intractable, we might look more closely at his suggestion that a demonstration of heritable differences in the distribution of abilities among different ethnic groups might be irreconcilable with human justice. Apart from the implication of a fixed rather than an adaptive concept of justice, whose problems I have noted above, this proposition seems to be blaming the messenger for the message. For science does not create the realities of nature: it only discovers them. And if it is not allowed to discover them they will still be there, determining whether or not our assumptions and our predictions turn out to be correct.

Recognition of the distinction between reality and the knowledge of reality has profound consequences. It tells us that if we wish to build social policy soundly we must not confuse the normative with the empirical. More specifically, we must rest the goal of racial justice on grounds of moral conviction, rather than on vulnerable assumptions about questions of fact; and we must recognize that we can adapt our social institutions to our evolutionary legacy, but not *vice versa.* We must also recognize that justice and equality are subtle and complex concepts, however simplistic the forms that they assume in the ideological marketplace: and these concepts will eventually have to be defined in ways that do not depend on a particular assumed distribution of abilities. If we choose otherwise, and suppress human behavioral genetics for fear that the results may contradict our assumptions, the costs may be high. For a major goal of this field, long emphasized by J. B. S. Haldane, is to help us to adjust educational procedures to individual differences in cognitive potentials and in patterns of learning.

For several reasons, then, the assumption of an inherent conflict between genetics (or other areas of science) and justice seems philosophically unsound. The objections can be summarized quite simply: since blocking off an area of inquiry on moral grounds fixes our knowledge in that area, it becomes, in effect, an illogical effort to derive an "is" from an "ought." I would suggest that we call this procedure the moralistic fallacy, since it is the mirror image of what David Hume and G. E. Moore identified as the naturalistic fallacy. But, alas, identification may not get us very far. For as Stephen Toulmin recently emphasized in *Daedalus*, we are in the midst of one of history's swings between a romantic concern with the good and a classic concern with truth.

4

Review of *The Double Helix*

James Watson's story of the discovery of the structure of DNA *not only was a* succès de scandale, *but it has justifiably become a classic. Nevertheless, when the book came out in 1968, I was disappointed to find that only one reviewer, Andre Lwoff, adequately distinguished the intellectual greatness from the moral deficiencies of Watson's performance, both in his conduct of the research and in writing the book. I would have been happy to try to reinforce that point of view, but no occasion arose. However, when the book was reprinted in 1980 in Norton's Critical Editions an invitation came—which was even better because it provided an opportunity to discuss the moral criteria of various reviewers, as well as of Watson.*

My review is placed here because it raises questions of moral philosophy. (They will appear again, in relation to evolution, in Part Two of this volume.) In addition, balancing my repeated emphasis on the value of objectivity for science, this review recognizes that in interpersonal relations too much objectivity, untempered by other, human considerations, is destructive. Lwoff expressed it particularly well in his review, as I note below.

The Double Helix generated controversy enough in 1968. Francis Crick was so outraged that he persuaded Harvard University not to lend the title its scholarly imprimatur, and he started (but soon dropped) a counter-book entitled *The Loose Screw. Nature* abandoned efforts, after twelve rejections, to find a reviewer among molecular biologists. But James Watson knew that a *succès de scandale* would have much more impact than a scholarly contribution to scientific history. How right he

---

American Scientist 70 (Jan.-Feb., 1982):76.

was: these confessions not only have sold over a million copies, but they now join *Faust, Crime and Punishment,* and Machiavelli's *The Prince* in a series of literary classics (though without the congenial company of Rousseau and Cellini). Even those who disapprove of some of the author's values will continue to be fascinated by this unvarnished account of the unorthodox, picaresque, yet fantastically determined way in which Watson and Crick got first to the discovery that was to revolutionize biology.

It is clearly useful to have the book now reprinted along with selected reviews, later comments by the other principals in the story, and the key scientific papers. In reviewing this version I shall comment primarily on the impact rather than the content of the initial publication, and on some of the added material. I shall assume that Watson would wish me to reciprocate his own blunt style.

On the positive side, the book has entertained and instructed a much larger audience than is usually concerned with scientific discovery, and it has no doubt inspired many students. Moreover, as Medawar's review emphasizes, it has disabused laymen of the misconception that science is simply a straightforward, inductive method for logically cranking out discoveries. For Watson's story illustrates vividly the importance of the creative imagination, coupled with critical testing against reality; and nowhere are the obstacles to successful creativity and the excitement of the search better seen. Indeed, it is still not obvious why it took two such very clever fellows months to think of obligatory base pairing, after they had accepted Rosalind Franklin's evidence for a multiple helix with the bases inside, knew Chargaff's rules for the base ratios, and were sympathetic to the idea of complementarity as the key to gene replication. And how obvious the model seemed, once it was imagined!

In another perceptive review, Robert Merton emphasizes that competition and concern over priority and credit are not to be deprecated but are central to the scientific enterprise. Yet he might have given more attention to the matter of degree. The image of a Nobel Prize, and the sense of a contest with other scientists (and not simply with Nature), played a larger role in Watson's motivation than in that of most scientists—including Crick (as is evident from Judson's later *Eighth Day of Creation*). Thus, while we have an honest account of a scientific discovery, it is also a highly unrepresentative account, almost a caricature in its psychology. Not only among the yeomen of science, but also among the Nobel Prize winners and potential candidates in my acquaintance, I find few for whom I could imagine some of the behavior described here —which does not mean that they have been unconcerned with credit.

The consequences of widespread attention to so atypical a picture have not all been salutary. One unfortunate effect, I believe, was on the public image of science, in an age of waning confidence. Innumerable reviews by literary critics (not reprinted here) gloatingly concluded that this book had finally let out the truth about scientists. These mysterious priests in the temple are evidently no more virtuous than businessmen or politicians, and thus their demands for public trust and for autonomy, based on the inherent objectivity of science, are not warranted.

The impact of the book on the values of young scientists has perhaps been more serious, for Watson has inevitably served as a role model. Again, other factors have contributed to the increasing competition and secrecy of our era—most recently and forcibly the possibility of literally getting rich quickly. But this book, with its contempt for old-fashioned restraints and its unabashed worship of success, laid the groundwork. . . .

I find disturbing Stent's ascription of "naiveté and self-righteousness" to those who criticized Watson on moral grounds—as though his great intellectual achievements exempt him from moral judgment. For me the most profound review was that of Lwoff, who described Watson (at the time of the discovery) as a person virtually without affectivity, and without awareness that the naked truth is a deadly weapon, to be used with discretion. It is remarkable that so many reviewers ignored this moral issue.

I am not suggesting that reviewers should, or could, dissuade Watson from behaving, whether because of an irresistible impulse or a calculated decision, as an *enfant terrible*. In their responses, however, they do affect the impact of his example on the upcoming generations of scientists, and thus they help to condition the kind of world we will live in tomorrow. To be sure, because of Watson's intense dedication to quality in science and his remarkable ability to recognize and inspire talent, the world has benefited very much from his contributions and easily forgives his faults. But could a scientific community function if such lack of concern for others' feelings became the norm?

Despite the emphasis on egocentricity, the story ends as a morality play. Franklin, Wilkins, and Pauling, on being shown the base-paired atomic model, are each described as instantly recognizing its significance and expressing genuine pleasure. Had one of them won the race, I am sure that Crick—and even Watson, despite his discouraging self-portrait—would have reacted in the same way.

5

Objectivity Versus Doctrine

Sir Peter Medawar published in Nature *a laudatory review of* Not in Our Genes, *a book that struck me as largely propaganda rather than science. Because he is such an influential scientist and writer it seemed worthwhile to question whether the authors' assertion of their humanitarian intentions justified his favorable review. It was gratifying to receive an appreciative response from Sir Peter. Part Three of this volume contains my own review of the book, with a fuller exposition of my objections.*

It has been fashionable for scholars on the periphery of science to attack the idea that science can be objective and value-free. This view stems from confusion between different meanings of the word *science:* as a highly subjective activity, a methodology that maximizes objectivity, and the resulting body of knowledge. Although we have become more aware that unconscious bias is often hard to avoid, the goal of objectivity remains the foundation on which the success of science rests. No less than in Galileo's day, the deliberate introduction of ideological preconceptions into the scientific process undermines its integrity.

I was therefore dismayed that Sir Peter Medawar reviewed *Not in Our Genes*, by Lewontin, Rose, and Kamin, with enthusiasm (*Nature* July 19, 1984, p. 225). Moreover, he even quoted approvingly part of the authors' frank statement of political purpose: "We share a commitment to the prospect of the creation of a more socially just—a socialist—society. And we recognize that a critical science is an integral part of the struggle to create that society, just as we also believe that the social function of much of today's science is to hinder the creation of that

---

Nature 311 (1984):294.

society by acting to preserve the interests of the dominant class, gender, and race." Unfortunately, in admiring the goal of a just society, Medawar overlooks the possibility that the authors' commitment to Marxist doctrine would conflict with their commitment to objectivity.

In fact, the authors skillfully distort the views of scientists interested in human behavioral genetics, intelligence testing, and sociobiology, condemning research in these areas as worthless rather than as necessarily limited in precision. Indeed, almost no biomedical research could meet their perfectionist standards. Moreover, they project upon their victims a political motivation equal in intensity, but opposite in direction, to their own. They even condemn Kety's classic demonstration of a major role of heredity in schizophrenia, ignoring the fact that genetic studies on this disease are likely to lead, through the reductionist molecular genetics that they decry, to specific chemical therapy.

Historically, this book must be regarded as part of the long campaign of a group from the radical left, called Science for the People, to outlaw the study of human behavioral genetics. While most of the book repeats this group's earlier arguments, there is one major shift. After many years of trying, with little success, to convince the world that genes have little to do with individual differences in behavior or in potential, Lewontin *et al.* now deny any such naive cultural determinism and adopt the position long held by their opponents: intelligence is the product of gene-environment interactions. But their turnabout seems to be more a matter of strategy than of conviction: note the title of the book, and the statement that J. B. S. Haldane and H. J. Muller "argued (along lines that we would not) that important aspects of human behavior were influenced by genes" (p. 73). Moreover, instead of gracefully seeking an end to the sterile polemics, they claim the high middle ground of interactionism for themselves, and they cast down their opponents with the epithet "biological determinist" repeated (in a familiar political tactic) on virtually every page.

The authors are unusual not only in the righteousness but also in the manners that they bring to scientific controversy. For example, after quoting Louis Agassiz's assertion (in the nineteenth century) that the human sciences can in principle be freed of politics and religion, they add that "The sentiment was echoed in 1975 by yet another Harvard professor and biological determinist, Bernard Davis, who assures us that 'neither religious nor political fervor can command the laws of nature.'" Then follows another quotation from Agassiz: "the brain of the negro is that of the imperfect brain of a seven-month infant in the womb of the white." As a reviewer has noted, this slur, passing three authors and an editor, says much about the intent of the book (see M. Konner, *Natural History*, August 1984, p. 66).

Why does such doctrinaire and ambiguous rhetoric appeal to many thoughtful people, as it has to Medawar? Obvious reasons include the past misuse of genetic theories to support racism, fear that genetic studies of behavioral traits might reveal differences between races, and the belief that it is racist even to entertain that possibility. But this belief, however well intentioned, is profoundly illogical: racism is the willingness to have race, rather than individual qualities, determine a person's social treatment. Moreover, modern genetics has made a major contribution to the struggle to overcome this evil. For the intellectual justification of racism has been the ancient assumption that the differences between races are typological—that all the members of one group differ from all the members of another. But in contrast to the naive support of this view by certain geneticists in the past, modern population genetics has utterly destroyed its typological foundation. We now know that the genetic differences between races (except for some physical traits subject to climatic selection) are statistical and overlapping, and so one cannot infer an individual's potential from his race.

Future advances in understanding human diversity could also help us to reach humanitarian goals in another way, by improving our efforts to develop individual potentials. To be sure, since knowledge of genetic differences, like almost any scientific knowledge, could be used for ill as well as for good, it is understandable that some people focus on the immediate danger. But the constructive social impact of the shift of genetics from typological to populational thinking encourages confidence that its further advances can have a similar effect.

It is sad to see one with Medawar's eminence, and with his credentials as a writer on the philosophy (and on the manners) of science, lending credibility to this book. He does criticize one feature, its attack on reductionists, because he finds that they simply do not exist as described. But he does not question the existence of biological determinists, even though they are equally imaginary among modern biologists. He also approves what he terms the "right-thinking" quality of the book. But "right-thinking" (or orthodoxy) is a dangerous concept in matters involving science. If it refers only to the general goal of a just society, or to Medawar's stated belief that the world is in need of change, fine—but if it encompasses or condones political bias in the evaluation of science, it presents a grave challenge.

The extreme positions and the political propaganda presented in *Not in Our Genes* are unlikely to influence the views of many scientists close to the field. Nevertheless, the book may convince lay readers, and it will surely have a distorting influence on the public image of science.

Part Two

Evolution: Sociobiology, Ethics, and Molecular Genetics

6

Evolution, Human Diversity, and Society

Evolutionary biology and genetics are the areas of science that have most challenged traditional views on the nature of man. Since many other scientists have valiantly defended the teaching of evolution against the current attacks, I have not been heavily involved in this conflict. (However, another selection, number 10, will offer what I hope is useful ammunition.) Instead, I have taken up the defense of a second social implication of evolution: a major role of genes, interacting with the environment, in human behavioral diversity. I address this controversy more directly in Part Three, but here I outline the scientific base for that discussion and then comment on a new branch of evolutionary biology, sociobiology.

Earlier biologists necessarily focused on the rich diversity of form and function in the living world, since visible features were all that they could observe. With the development of techniques for studying organisms at a cellular and a molecular level, experimental biologists shifted their emphasis increasingly to the other face of evolution: the unity that underlies this diversity. However, with the powerful tools that molecular biology now provides, we can soon expect exploration of the universal features, shared by all organisms from bacteria to man, to be fairly complete. The main focus of biology will then inevitably shift again to diversity, both between and within species.

Modified from Zygon 11, no. 2 (June 1976):80-95. Presented at the Twenty-second Summer Conference ("Genetics, Biological Evolution, Ethics") of the Institute on Religion in an Age of Science, Star Island, New Hampshire, July 26, 1975.

Paradoxically, at the same time that this flowering of biology has been giving us such deep insights, as well as rapidly expanding applications in medicine and agriculture, we have also seen the rise of widespread public disenchantment with science, for many reasons.[1] Genetics in particular has become a source of anxiety. I have discussed elsewhere one major area of concern: genetic engineering.[2] The present paper deals with another aspect of genetics that raises much deeper philosophical, religious, and political questions: the implications of evolution and genetics for the nature of our species, and the relevance of this knowledge for the concepts of equality and social justice.

Darwinian Evolution

Darwin's theory of evolution by natural selection is a unique product of the scientific method, for it was originally based almost entirely on historical inferences rather than on hypotheses validated by experimental tests. Nevertheless, it is one of the greatest triumphs of that method, and it is clearly the most important generalization in biology, accounting for both unity and diversity in the living world. Moreover, its realistic picture of man's origin accounts for his unique qualities as a pinnacle in evolution, replacing earlier speculations that tried to account for these qualities in other ways.

The theory of evolution required an unexpectedly long time scale. Its dates are now based not only on evidence from geological structures but also on the much more direct evidence from the decay of radioactivity. Hence there is unlikely to be any major correction in the current estimates. To review these briefly: Life on earth began as one-celled organisms about three billion years ago. Only in the last 1/1,000 of the total period of evolution, about three million years ago, did the hominid line, leading to man, branch off from the other apes. Man's rapid cultural evolution, using the written word to accumulate information and agriculture to accumulate surplus goods, occupied about 1/500,000 of the total—about the last six thousand years. Only in the last 1/20 of that period have we had the scientific method, in which verifiable evidence and testable hypotheses supplement pure reason in our efforts to deepen commonsense understanding of the natural world. And in those three hundred years it is only a bit more than one hundred since Darwin's *The Origin of the Species* was published and only forty since Oswald Avery identified the material substance of the gene. The ethical implications of evolutionary genetics are thus very new, and it is not surprising that we are having trouble elucidating and assimilating them.

When Darwin finally published *Origin* in 1859, after incubating the theory for over two decades, he stopped short of discussing the implications for man's origin, though they were clear. Only ten years later did he develop the courage, after watching the intense intellectual controversy that he had precipitated, to spell out this final conclusion in *The Descent of Man*. It was vigorously opposed by the religious establishment on the grounds that the idea of the evolution of man from lower animals by natural selection destroyed the foundations of public morality. In addition, the whole theme of evolution and change was anathema to a social establishment dedicated to preservation of the status quo.

The polemics of the mid-nineteenth century dwindled after a few decades, but by no means with a clear victory for the Darwinians. The scientific evidence was not complete enough to overcome skepticism, and even many biologists remained unconvinced until about the 1930s. For natural selection could not occur unless the hereditary process produced stable new variations for selection to act on. But since evolution was discovered before the elements of genetics (even though the reverse sequence would have been more logical), Darwin knew nothing about the mechanism of heredity, with its double property of constancy and variation.

In fact, it was only five years after *Origin* that Gregor Mendel, abbott of a monastery at Brunn, discovered the existence of fixed, independently segregating units of inheritance, each governing a specific trait. But his statistical approach was foreign to biologists of the time, and so the work was buried. It was rediscovered in 1900, and by then some biologists had become receptive to this new mode of analysis, whose value in physics and chemistry had already become evident.

I emphasize this point because the teaching of elementary mathematics in our educational system still does not include a grounding in the fundamental concepts of statistics. Moreover, nonstatistical, qualitative thinking is built into the very structure of our language, so discussions that involve statistical concepts frequently result in failure of true communication. For example, in discussing genetic differences among people, if I say that group A is better endowed than group B in some respect, I would have in mind distribution curves whose mean values differ. I thus take it for granted that this generalization tells us nothing about the standing of any specific members of either group. However, you may think you hear me suggesting that all members of group A are better endowed than any members of group B. This misunderstanding has been the source of enormous confusion, mischief, and polarization.

The Synthesis of Evolution and Genetics

For several decades the field of genetics remained quite separate from evolution. In the kinds of traits that Mendel and his early successors dealt with a single gene determines a specific trait—say, blue or brown eyes, or one or another blood group. Moreover, in most organisms every gene is present in two copies, which may be identical or may be different; and in sexual reproduction an offspring receives from each parent, more or less randomly, one member of each of that parent's gene pairs. This reassortment, along with the dominance of one form of the gene over an alternative, recessive form, determines the visible trait (phenotype).

Evolutionists, however, are interested primarily in those morphological and behavioral traists that vary in a quantitatively continuous manner, rather than existing in only two alternative forms. These traits did not seem to be inherited according to Mendel's laws, and it took several decades to work out the statistical demonstration that their genes also obey those laws. The difference is that the qualitative traits are polygenic rather than monogenic—that is, a large number of genes contribute to the intensity of a trait, and the variety of their combinations gives rise to an apparently continuous range of values.

Before genetics could be effectively applied to evolution, a second concept had also to be clarified: the interaction of genes and environment. Mendel's studies—with plants growing in a relatively uniform environment—emphasized the deterministic effect of the genotype (the total set of genes in an individual) on the phenotype (the set of traits observed). But we now know that only a few of the traits that we observe are determined in this way. With most, genes determine the range of potential of an individual, and within that range the interactions with the environment condition the actual phenotype that develops. For example, we know that tall parents tend to have tall children and short parents tend to have short children. However, since the mean height of college students has been increasing over the past seventy-five years, it is also obvious that differences in nutrition (and perhaps in other environmental factors) can affect height. If it seemed socially desirable we could attempt to equalize stature by giving optimal diets to the children from short families and poor diets to the children from tall families. But the success of this form of egalitarianism would still be limited by the ranges of genetic potential of the individuals—and where these ranges did not overlap one could not achieve equality, though one could decrease differences.

Modern molecular genetics has reinforced and explained the mechanisms underlying these principles of classical genetics. We now know that some genes are simply structural, determining the structure of a corresponding protein. When a trait is determined by the nature of that protein (e.g., normal versus sickle-cell hemoglobin) or by its absence (e.g., absence of a pigment-forming enzyme in albinos), the trait is strictly Mendelian and monogenic in its inheritance. Other traits, however, whose intensity depends on the environment, involve regulatory genes: genes whose protein products sense appropriate environmental stimuli and thus influence the activity of the genes that they regulate.

This mechanism was first established with simple bacterial cells, which respond to chemical stimuli in the environment. For example, the colon bacillus (a major inhabitant of our gut) can utilize the sugar lactose as a food, but in its absence the bacterial cells do not make the specific proteins that are necessary for that utilization. When lactose is added it complexes with a specific regulatory protein, and this complex activates the genes that make the proteins required for utilizing lactose. Moreover, mutations in the regulatory genes alter responsiveness to lactose. In humans similar cellular responses to specific chemicals have been observed. And there is no doubt that other kinds of stimuli, perceived by our sense organs, are ultimately translated (through the mediation of the nervous system) into chemical stimuli that either activate or repress specific genes in appropriate cells. Differences in regulatory responses no doubt are the major molecular basis for individuality.

This knowledge from molecular genetics is certainly relevant for our understanding of intelligence and of other mental traits, but only in general terms: many genes must affect intelligence; they act through the production of proteins that ultimately influence both the wiring diagram of the ten billion cells of a human brain and the functional properties of their connections; various of these genes must differ from one person to another; and the function of the switches (and in early life the formation of these connections) is markedly influenced by learning experiences. But so far we can deal with these genes only in the formal terms of the analysis of polygenic inheritance and not in molecular terms.

Molecular genetics has also provided extremely direct evidence for evolution—far more direct than the stepwise morphological variations and the homologies (in different living species, in the fossil record, and in embryological development) that led Darwin to his brilliant synthesis. Hence today one cannot rationally deny human evolution and at the same time accept the validity of science as the means of understanding the world of nature—a validity that each of us confirms innumerable times each day in using the fruits of technology. Indeed, I would say that Darwin's theory is now more than a mere theory. It is as firm a law as Newton's laws of motion or the laws of thermodynamics (though its implications are less fully understood).

Nevertheless, it is easy to see why many people still fear that the replacement of special creation by evolution threatens the foundations of public morality. On the other hand, some of us believe that a deeper exploration of the social implications of our knowledge of evolution may even help to provide a firmer foundation for our moral values. The rest of this paper will consider these two opposing views.

Misapplications of Genetics and Evolution

There is unfortunately a real historical basis for fear of efforts to relate evolutionary theory to society, for early efforts at such extrapolations not only were unsound but had tragic consequences. The first of these efforts, named "Social Darwinism" (but really the product of Herbert Spencer), focused exclusively on the role of competition in natural selection. The resulting exposition of an alleged natural law was used widely to rationalize the exploitation and cruelties of unrestrained laissez-faire economics. Only many decades later was it recognized that the evolution of social species, ranging from insects to man, has also selected for cooperation. Moreover, kinship selection can now explain the evolution of even an instinct (or a willingness) for altruistic self-sacrifice: The sacrifice of an individual can promote the spread of his genes if it aids the survival, and hence the multiplication, of kin who bear the same genes.[3]

Another premature application of genetic ideas was eugenics. Sir Francis Galton, a cousin of Darwin, advocated such a program, with the aim of improving the stock of our species just as animal breeders had improved the strains of domesticated animals. But he vastly overestimated the role of inheritance, compared with the role of favorable circumstances, in the achievements of the upper-class Britons whom he admired. He also greatly underestimated the cultural value of diversity. It is profitable to try to maximize an obviously valuable trait in domestic animals, such as speed in a race horse or milk production in cows; but in man our goals are not so simple, and there is no self-evident ideal to select for.

Unfortunately, both the eugenic movement and Social Darwinism were used to bolster ancient notions of racial superiority and inferiority. These misapplications of genetic ideas contributed to the restrictive immigration laws of 1924 in this country, and they reached their culmination in the Nazi idea of the master race and its right to engage in genocide. But modern population genetics, as I have noted elsewhere, has radically revised our concept of race, and in so doing it has thoroughly dispelled the prescientific assumptions and the pseudoscientific rationalizations that perpetuated these pernicious social views.

We also now know that neither the 100 percent hereditarian view nor the 100 percent environmentalist view of human behavior can be defended. Both genes and environment contribute to the observed variation in a population, and their relative contribution will vary from one trait to another. Moreover, this proportion, often expressed quantitatively as heritability (the ratio of genetic variance to total variance), will also differ from one population to another, depending on the distribution of its genes and its environments.

Heritability can be measured in experimental animals in two ways: by exposing a variety of genotypes to the same environment, or by exposing the same genotypes to a variety of environments. Since we cannot control these variables as completely in man as in experimental animals, the numbers obtained have a much larger margin of error. But there is no doubt that genes and environment both contribute a good deal to such traits as, say, general intelligence. Nevertheless, people interested in advancing our knowledge in this field are sometimes accused of being biological determinists, perpetuating obsolete nineteenth-century dogmas. One might as justifiably identify a modern surgeon with the phlebotomists of past centuries!

Implications of Evolution for Human Genetic Diversity

Let me further emphasize that, even if no one had ever devised a test for measuring IQ, we could still be confident, on grounds of evolutionary theory, that our species contains wide genetic variance in intelligence.[4] The reason is that natural selection cannot proceed unless it has genetic diversity, within a species, to act on; and when our species is compared with its nearest primate relatives, it is obvious that our main selection pressure has been for an increase in intelligence. Indeed, this change proceeded at an unprecedented rate (on an evolutionary time scale): in the past three million years the brain size of the hominid line increased threefold. Yet this period is so short that our DNA as a whole changed by only 1 percent from that of our present nearest primate relatives, and our biochemical traits changed little; moreover, the changes in our physical traits were mostly those subject to the same selection pressures as intelligence, because they made it more useful (e.g., opposable thumb for making and using tools, bipedal posture to free the hands, a female pelvis with a larger birth canal to accommodate a larger cranium). It is as though once the trick of abstract thought emerged in evolution it had such selective advantage that it was intensified at a remarkably high rate. Such rapid selection for increased intelligence could not have occurred unless the selection pressure had a large substrate of genetic variation to act on.

We may also note that the uniqueness of man arose from this pressure for rapidly intensifying the valuable, novel traits of the hominid line, which increased its capacity to adapt to novel circumstances and to manipulate the environment. The result was that a single hominid species emerged to populate the whole earth, whereas other families of organisms have numerous species, at similar levels of neural development, occupying different ecological niches.

Clearly, then, evolving mankind must have had a wide range of genes that affected behavioral traits. To be sure, these traits exhibit unusually great plasticity of response to the environment, so their genetic components are difficult to measure. For this reason, reinforced by emphasis on cultural evolution, some anthropologists have suggested that in our species cultural adaptability has replaced genetic diversity. But this is a fanciful concept. Such a dramatic switch from recent, great biological variation to present virtual homogeneity would contradict all we know about the mechanisms of population variation and the slow pace of evolution. There is every reason to believe, from first principles, that mankind is still evolving.[5] Moreover, since our species still possesses a large, easily demonstrable reservoir of genetic variation for both physical and biochemical traits, and since our behavioral traits have evolved even more rapidly, I would find it impossible to entertain serious doubts that these traits also have such a reservoir.

We see widespread reluctance to accept this concept today, based on fear that it will undermine the struggle for greater equality. Indeed, one of the implications of evolution, as noted above, is that long-separated populations, subject to the pressures of different environments, will accumulate statistical differences in genes that affect behavioral potentials, just as in their other genes. Evolution does not predict the magnitude or even the direction of such differences, but it does say that we cannot predict the numerical outcome of mental testing if barriers to equality of opportunity are removed.

This is a painful message for those who are deeply concerned with social justice, and I wish we did not have to face it. But if we wish to pursue the goal of equality on a realistic basis we must recognize the fundamental difference between social equality, which we can legislate, and biological equality or inequality, which is beyond our control. If we insist on assuming a nonexistent biological equality between people we will pay a large price in the long run. Thus if we set unattainable goals in education we will demoralize our teachers by blaming them for every failure, and we will thrash about from one program to another because none reaches the assigned goals. We will ensure chronic social unrest by promoting a profound fallacy: That because unequal achievements have often been due to unequal opportunities (which is true) they are proof of unequal opportunities (which is false). We will promote guilt and friction among parents by making them consider their faulty guidance responsible for all behavioral problems in their children. And we will jeopardize the struggle for racial justice by basing it on fragile, conceivably disprovable assumptions about matters of empirical fact (the distribution of potentials) rather than on moral and political convictions. On the other hand, the better we can identify differences in various potentials, and in patterns and rates of learning, the better we will be able to provide true equality of educational opportunity—that is, the opportunity to have everyone's education equally designed for maximal fulfillment of his/her potentials.

If equality of opportunity, combined with the existence of genetic heterogeneity, produces a result that does not satisfy society's strong pressure for greater equality of outcome, biological considerations suggest that we should examine more closely what we mean by equality of outcome. At present we seem to be aiming at leaving the reward system more or less untouched, and instead trying to satisfy the social pressures by setting up quotas for distributing the more highly paid or prestigious jobs among various identifiable groups. This solution seems unstable to a biologist, compared with an economic rather than a vocational egalitarianism—one that would aim at matching responsibilities with abilities but would then increase equality in the reward system.

It is ironic that recognition of genetic diversity as an implication of evolution finds intense opposition from the Left today, just as the implications of evolution with respect to our origins aroused opposition from the Right a century ago. Yet a pluralistic society should be able to recognize our biological diversity as a great cultural asset. Indeed, just as our rapid biological evolution required a wide range of variation for natural selection to act on, so our rapid cultural evolution depends on the capacity of the population to generate, and then to select in its social practices, from a variety of behavioral responses to new challenges; and that variety in response obviously has been enormously increased by our variety of genetically conditioned potentials, drives, and preferences. Indeed, if nature had selected for behavioral genetic homogeneity in our species, or if we should set up a successful eugenic program with this ultimate egalitarian goal, then it is clear that even if we selected the most admirable traits we would have a much duller culture. We would also decrease the adaptability of our species to unforeseeable changes in the environment—a property of the utmost importance for our survival.

I would further suggest that the polemics over the heritability of IQ not only have blinded us to the advantages of diversity but have seriously distorted our perspective. The very intensity of the opposition fortifies the tendency to treat IQ measurements as an index of human worth, rather than as a useful index of likely performance in certain types of education. Instead of fervently denying the existence of genetic variation in intellectual potentials, or the practical value of tests as guides for educational placement, it would be much more constructive to admit and to emphasize the existence of multiple kinds of intelligence, the real but limited social significance of differences in intelligence, the value of many other traits, and the cultural value of diversity.

Evolution and Ethics

I would now like to discuss some aspects of the interaction of science, and particularly of evolution, with the problems of morals. In the nineteenth century this interaction led to a war between science and theology, based on fear that public morality would suffer if we abandoned the transcendental, metaphysical conceptions that had long provided its foundation for a majority of people in the Western world. From the point of view of a scientist, established religion was wrong in the position it adopted, for it was led to oppose verifiable truths about the world of nature, and it was bound to lose. But we can now see that the clergy were right in their prediction of troubles ahead. Since an increasing fraction of the population can no longer accept traditional, supernatural explanations for the origin of a moral code, the public moral consensus has been attenuated. This development has no doubt contributed to the weakening of social bonds and to the recurrence of barbarism in enlightened societies.

But while the conflict between science and religion is far from resolved, recent advances in our understanding offer promise of helping by eliminating some misconceptions that have clouded the issues. First, I would emphasize that scientism—the assumption that science can solve any problem—is obsolete. We are only now recovering from this illusion. At the same time, we must recognize that science is not irrelevant to problems that involve values. For in choosing a goal we not only make a value judgment but also estimate the relative feasibility and the consequences of alternative goals. Science can help us make those estimates more realistic and reliable. The scientific method for understanding the world of nature, and the concern of religion with goals and values, can thus be viewed as complementary guides to action rather than as conflicting approaches trying to take over each other's territory.

A second advance is the increasing sophistication in our understanding of the simultaneous evolutionary selection of competitiveness and cooperativity. Sociobiology has now accounted for even the evolution of extreme altruism, leading to self-sacrifice for the common good. Moreover, even with disease-producing viruses and bacteria, a strain that rapidly kills off its host is not as successful (i.e., does not multiply as much) as one that can multiply for a long time within a surviving host and thus has more time to infect another host.

The conflict between cooperative and competitive drives in man thus is not unique but is an example of the usual ambivalence of evolution, selecting for balance and compromise between opposing traits. Religious leaders have long recognized this duality as an inherent feature of the human condition, and Freud described it in terms of superego and id, or eros and thanatos. Sociobiology now provides an additional approach to the problem, deeply embedded in reality and aiming at the modest, but solid, stepwise advances characteristic of science. And just as the uncertainty principle in physics has helped to illuminate the nature of matter, so a recognition of the biological roots of conflict, and the limitations in our power to eliminate it, may help us to set realistic goals and to identify the factors that we can control profitably.

Finally, I would suggest that we should reevaluate one presumed implication of evolution that has had particularly destructive consequences: the view that eliminating the traditional absolutist framework for ethics necessarily leads us to the alternative of complete moral relativism, in which anything goes. In the light of sociobiology this is not a logical conclusion. For since evolution has built into every kind of organism a deep-seated drive for survival of its species, and since we have evolved as a highly social animal, we must have within us strong, genetically determined instincts for patterns of social behavior that are compatible with that survival. Our evolutionary endowment thus is incompatible with unlimited moral relativism. It requires restraints on our behavior, based not only on self-interest but also on an instinctive interest in the welfare of our group and our progeny.

Language may provide a useful model for our genetic predisposition to social behavior. We are not born with a particular language, but we are born with the capacity for learning a language; and while our cultural evolution has created many languages, which differ enormously in detail, they all have deep structural features in common. As Chomsky has emphasized, these common features must reflect anatomical structures in the parts of our brain that are concerned with language.[6] A student of evolution would add a thought about origins: The needed structures are there, and the language that they use corresponds closely to various aspects of the world around us, only because those structures have evolved in response to the pressure to communicate with one another in increasing detail about the world around us. We could not transform sense perceptions, and novel associations of remembered perceptions, into a vocabulary of thousands of words unless our genes had built into our brain the required sets of connections, which are there waiting to perform those tasks. Similarly, we are not born with a detailed ethical system, but we are born with the capacity and the need to develop an ethical system, whose details will vary, like those of language, from one culture to another.

Sociobiology thus contradicts the arguments for extreme moral relativism. In so doing it provides a biological base for the insights of the ancient religions, and for the traditional and universal aims of education, parental guidance, and psychiatry: to help people balance immediate gratifications with long-term goals, and aggression with love. It does not deny the role of ritual and emotional appeal in reaffirming and strengthening recognition of nonhedonistic moral values. Instead, it complements religion by substituting a realistic base for one that is no longer plausible for many people. Moreover, by recognizing species survival, and not individual survival, as the overriding biological goal, sociobiology can help us to define the range of values compatible with this survival. It may thus supplement traditional approaches to our truly novel and frightening ethical problems, which are being generated by our accelerating alterations of the world around us, by the expanding communications among people in all parts of the earth, and by the increasing ratio of population to resources.

Let me close by reemphasizing a value that is especially dear to scientists: the habit of truth.[7] Experience has taught scientists that in their area (in contrast to many other human activities) distortion of the facts does not pay, for nature always has the last word. The same value is also relevant for the problem of achieving a more just society. For while this problem is not primarily a scientific one, the success or failure of our approach will depend in part on the correctness of its underlying assumptions about the facts of human diversity. And here nature will again have the last word.

NOTES AND REFERENCES

1. B. D. Davis, "Novel Pressures on the Advance of Science," *Annals of the New York Academy of Sciences* 265 (1976): 193–202.
2. B. D. Davis, "Prospects for Genetic Intervention in Man," *Science* 170 (1970): 1279–83.
3. E. O. Wilson, *Sociobiology* (Cambridge, Mass.: Harvard University Press, 1975).
4. B. D. Davis and P. Flaherty, eds., *Human Diversity: Its Causes and Social Significance* (Cambridge, Mass.: Ballinger Publishing Co., 1976).
5. T. Dobzhansky, *Mankind Evolving* (New Haven, Conn.: Yale University Press, 1962).
6. N. Chomsky, *Aspects of the Theory of Syntax* (Cambridge, Mass.: M.I.T. Press, 1965).
7. J. Bronowski, *Science and Human Values* (New York: Harper Torchbooks, 1956).

7

The Sociobiology Debate

The publication of Edward O. Wilson's Sociobiology *in 1975 synthesized a great deal of material and essentially created a new field. The book was bound to raise hackles, with its challenging and dramatic prediction that this field would eventually cannibalize the social sciences and the humanities. However, it is unfortunate that Richard Lewontin and his disciples launched a virulent and personal attack, accusing Wilson of racist and reactionary views that simply were not there. The poisoned atmosphere created by this attack delayed by at least a year most of the serious scholarly analyses and criticisms that the book merited and eventually received.*

The following published letter is a product of the early part of that period, in which I was interested in defending Wilson, and the study of the biological basis of social behavior, against ideological attacks. The next piece in this volume will present my own detailed criticisms of certain of Wilson's views.

Arthur Caplan's thoughtful review of E. O. Wilson's *Sociobiology* discusses the problem of evolution and ethics from the point of view of philosophy, which aims at creating a comprehensive system of thought based on clearly defined assumptions and logically consistent inferences. Science, in contrast, aims at understanding the concrete realities of nature, and in its step-by-step approach it is more concerned with the solidity than with the comprehensiveness of its growing body of knowledge. As we grope today for constructive interactions between the two

Hastings Center Report (October 1976):19.

approaches, we must recognize that sociobiology can make only limited contributions toward solving social problems. If we expect it to provide "the" basis for an ethical system, we will surely be disappointed, and we may miss what it can provide.

I would therefore like to comment on Caplan's statement that Wilson seems "merely" to urge a new inquiry into the issue of ethical naturalism, without being committed to a belief in evolutionary ethics. This is correct, if one accepts Caplan's definition of evolutionary ethics as the search for an evolutionary foundation for specific ethical presciptions and value systems. But while such a deterministic formulation helps sharpen the issue, it also suggests the outmoded view that science can solve problems involving value judgments. We must reject such scientism. At the same time, we need not accept the opposite extreme—that is, the view that science is irrelevant to such problems. We can take an intermediate position: that even though science cannot solve normative problems, which are in principle incapable of an objectively correct solution, it can nevertheless make valuable contributions. Thus the scientific method can help us to evaluate the means and the consequences of reaching various goals, and these analyses can contribute, along with our value judgments, to our choice of goals. In addition, scientific insights into human nature are surely relevant to the problem of formulating ethical systems: for however wide the range of possible systems, it is clear that any viable system must be consonant with human nature. In other words, our biological evolution has set broad constraints on our behavior, and between these borders our cultural evolution steers our course.

Wilson clearly is committed to this view. More specifically, he reformulates the ancient concept of human nature in terms of behavioral motivators and censors inherited from earlier stages in our evolution, and he anticipates that recognition of these factors will make our approach to problems of social behavior more realistic and effective. To be sure, he also goes farther and suggests that eventually a complete neuronal understanding of the human brain will provide a firm foundation for ethics. Such long-range speculation is clearly vulnerable, but it also does not seem to me very important, for it is hardly relevant to the real problem of exploring the ethical implications of sociobiology for the present and the foreseeable future.

It would be presumptuous to try here to specify these implications in any detail: that is one of the main jobs ahead. However, certain directions seem evident. The most general one involves the problem of a moral consensus. Ever since evolution undermined the transcendental foundation for ethics, which had guided the bulk of people in the West for millenia, many social critics have concluded that the only logical alternative is unlimited moral relativism. The consequences of this loss of a moral consensus have been disastrous. Humanity desperately needs a replacement that will appeal to enough people to restore an effective consensus. Philosophers have not been notably successful, for while they have continued their ancient search for a basis for ethics that would satisfy their criteria, their logical speculation, by itself, leaves too wide a range to lead to a consensus.

Sociobiology may help to bring us closer to the goal, thereby atoning for the earlier contribution of the theory of evolution to excessive moral relativism and to Social Darwinism. Thus the very history and structure of our DNA commits us to the primary evolutionary drive for species survival and adaptation. With this source of constraints we can restrict the *range* of acceptable ethical systems, in broad terms, to those that are consistent with this long-term goal. Sociobiology also provides a base for escaping from the shallowness of pure egoistic utilitarianism. Our brains are programmed for ethics, just as for learning and for language: in each area the specifics that we develop can vary enormously, but common features are built in. Hence philosophers need not apologize if they find intuitionist approaches useful. Finally, one might also suggest, but with less confidence, that the evolutionary drive for improved adaptation to the environment provides support for a perfectionist ethics, emphasizing excellence in those talents that give man his unique position (and his evolutionary success in spreading into the widest possible range of environments).

When we go beyond these very general implications of sociobiology and seek help in assessing more specific features of ethical systems our vision becomes more clouded. Wilson seems to anticipate that our increased understanding of the biological roots of human motivation and feeling will begin to be useful in this respect in the near future. I am less sanguine. But even if we cannot specify the relevant biological factors in enough detail to let them serve as a guide to policy, recognition of their existence can influence our viewpoint, if only by leading us to recognize limits to our social goals. For example, if there is a genetic component (as well as a socially conditioned component) to our competitive and our filial drives, to the rational and the irrational aspects of our behavior, and to our individual differences in drives, abilities, and tastes, we cannot hope to eliminate conflicts. We can hope only to moderate and contain them, and to achieve a reasonable balance between altruism and aggression. This aspect of the human condition has always been recognized by traditional moralists, but more recently it has been opposed by utopians imbued with unlimited confidence in the power of environmental manipulation. In closely related developments the pressure for increased social equality must somehow be reconciled with the wide genetic diversity built into our species, and with the strong tendency of members of social species to advance the interests of their close relatives (kin selection). The evidence from sociobiology on these issues may help to keep us on a realistic path.

Another interesting feature of Caplan's review is his subdivision of the subject of evolution and ethics into three parts: the evolution of ethics, evolutionary ethics, and the ethics of evolution. He expresses disappointment that Wilson failed to make clear exactly which of these value problems compel his interest in sociobiology. But a biologist might see evolutionary ethics (concerned with present problems) as simply a continuation of the evolution of ethics (concerned with origins): he would analyze both in the same Darwinian terms. As to the third category, the ethics of evolution: if the question concerns the past, any effort to pass judgment on what has been good or bad (in an ethical sense) about evolution, or about the properties of any species, would seem silly. Evolution is simply there, and ethical concepts cannot apply when there are no options under human control. If, on the other hand, Caplan is referring to conscious control over future human evolution (that is, eugenics) he is reaching into an area that Wilson has wisely considered to be outside the scope of sociobiology.

Finally, I would like to comment on Caplan's own possible moral relativism, in referring to "a harsh critical response from those reviewers sensitive to the inherent ethical difficulties," and in considering both sides of the debate equally impassioned. His safely tolerant position seems to me to miss an important philosophical issue. For while there is no reason to doubt the sincerity of this group of critics ("Science for the People"), we would be naive not to recognize that their criticisms are based primarily on ideological convictions, rather than on considerations of ethical sensitivity (in the usual, nonpolitical sense) or considerations of scientific validity. This group rejects claims that studies on the genetic and evolutionary aspects of human behavior can help us to find the roots of social problems, and can help us to optimize environments for different individuals. Instead they are convinced that such studies will impede efforts to bring about social changes that they consider desirable: hence they discourage these studies as vigorously as possible, using means that go outside the usual range of scholarly criticism (see article by Nicholas Wade in *Science*, March 19, 1976).

Within the political and philosophical framework of our culture, espousal of radical political or economic views is a precious right. But intolerance of intellectual freedom is another matter. When its scale is small enough it can be tolerated, but exposure still seems more appropriate than legitimation.

8

Review of *Sociobiology: The New Synthesis*

This review of Edward O. Wilson's famous book (Harvard University Press, 1975) is somewhat less critical than the essay that follows, perhaps in part because the ideological attacks of Science for the People on the book created a polarized atmosphere that for a time inhibited more balanced criticisms. Most of the material in this selection is expanded in the essay that immediately follows it.

In recent decades the behavioral and social sciences have abandoned an earlier interest in genetic aspects of human behavior and have focused almost exclusively on environmental factors. One reason is that the necessarily imprecise and indirect methods of evolutionary and behavioral genetics have contrasted with the spectacular methodologic advances in other branches of biology. However, the results of advances in molecular genetics will now surely promote public acceptance of man's evolutionary origin and its implications. Quantitative comparisons of DNA sequences have provided direct evidence for the evolutionary continuity of the living world, and detailed analyses of the mechanisms of hereditary continuity and variation have filled in the major gaps in the evidence for Darwin's theory.

The time thus seems ripe for this book, whose aim is to identify the evolutionary roots of social behavior through a rigorous, systematic approach, based on tests of predictions, not on "philosophical retrospection." The field would include the ecologic pressures that have selected for various patterns of social behavior, the relevant parameters in population genetics, and the adaptive or maladaptive function of traits

---

New England Journal of Medicine 283(1975):1375.

based on genes inherited from an earlier era. "When the same parameters and quantitative theory are used to analyze both termite colonies and troops of rhesus macaques, we will have a unified science of sociobiology."

The author has made a remarkably successful start in establishing this new field. Indeed, the bold and comprehensive nature of the volume reminds one of *The Origin of Species.* Observations collected from the entire animal kingdom are organized into a broad conceptual synthesis, and generalizations are supported by numerous specific examples. The 600 double-size pages incorporate material from over 2000 references, mostly in population genetics, ecology, and ethology. (Even information theory is included in some detail, for the essential feature of a society is "reciprocal communication of a cooperative nature.") The style is informed with flashes of irony and with loving care for language, and the pointilist drawings and novel format add to the book's attractions. However, the general reader will face a large task, for though the compact introductory chapters cover some of the important principles, others are left to the later, detailed exposition of evolutionary and social mechanisms. A synoptic version, therefore, would surely find a wide audience.

The author is not diffident in criticizing the social sciences. In his view their empirical correlations of particular phenomena with features of the environment resemble the earlier, purely descriptive phases of taxonomy and ecology, before these fields were integrated into a neo-Darwinian theory that weighs each phenomenon for its adaptive implications. The social sciences are seen as the last branches of biology waiting to be included in this synthesis. Moreover, ethology and comparative psychology, generally seen as the central, unifying fields of behavioral biology, "are destined to be cannibalized by neurophysiology and sensory physiology from one end and sociobiology and behavioral ecology from the other."

Reviews have so far not produced widespread protests against the threat of being cannibalized. However, the book has been accused of reinvigorating Social Darwinism. I find this charge grossly misleading. The term Social Darwinism has a well-defined historical meaning: a premature extrapolation from organic to social evolution, whose rationalization of political and economic exploitation was based on the confusion of an analogy with a scientific law and on the oversimplification of the evolutionary process as selection only for aggressive competition ("Nature red in tooth and claw"). In fact, *Sociobiology* emphasizes quite a different set of themes: the essential role of cooperation in social behavior, the discovery of mechanisms (group selection and kin selection) that account for its evolution, and the inevitable interplay of cooperation and aggression in all advanced social species, not only varying with species and with individuals but also responding, in predictable fashion, to specific environmental pressures and to the size and kinship of the interacting group. To equate this rich intellectual mine with the poverty of Social Darwinism is no more justifiable than to equate modern medicine with phlebotomy.

I find little to criticize in this superb volume. Though Wilson emphasizes the predictive value of mathematical formulations, he might still have acknowledged the more general contributions of such pioneers of neo-Darwinism as Dobzhansky. Mayr, Simpson, and Haldane. And though the ultimate aim is continuity between social and biologic sciences, the author may be underestimating the possibility of establishing important generalizations within a single level of organization—for example, Sherrington's insights into the integrative action of the nervous system did not depend on reduction to the level of cells or of molecules.

The largest problems appear in the last chapter, on man. In calling attention to the evolutionary aspects of such diverse human attributes as language, ethics, and war, and in aiming at maximal generality, the author engages in extraordinarily long-range predictions. Thus, "The transition from purely phenomenological to fundamental theory in sociology must await a full, neuronal explanation of the human brain," and "A genetically accurate and hence completely fair code of ethics must wait for the new neurobiology." At first glance such statements resemble the optimistic scientism of an earlier era, which expected the scientific method to solve not only problems about the objective world of nature but also human problems that involve subjective values. On the other hand, it would be rash to try to predict how far neurobiology—applying the scientific method not directly to questions of values but to the structures in us that create them—can eventually encompass the subtleties of human behavior. Since the book as a whole exhibits an extraordinary intelligence and perspective, Wilson's ultimate vision may well be simply extremely farsighted.

Whether this vision turns out to be hyperbole or distant reality, it should not hinder readers from recognizing the value of an evolutionary perspective in clarifying more modest aims: for instance, to recognize the problem of reconciling the goals of social equality and equity with the existence of wide biologic diversity in the human species, and to regulate aggression rather than to try to eliminate it. Above all, people have feared that evolution, in displacing earlier transcendental foundations for a moral consensus, must lead to unlimited moral relativism. But evolution can also provide an alternative foundation, for the deep-seated drive for perpetuation of one's genes, and the inescapably social nature of the species, places boundaries on the range of viable ethical principles.

None of these ideas, of course, are entirely new—Freud recognized them all. What sociobiology provides is a firm scientific basis for selecting among alternative general views of human nature. Moreover, the need is increasing. Novel ethical and political problems crowd upon us as a result of alterations in such sociobiologic parameters as population density, availability of resources, and communication between neighboring groups. And as Wilson emphasizes, these problems are being faced with a limbic system inherited from the Pleistocene.

9

Sociobiology, Human Individuality, and Religion

I immensely admire E. O. Wilson's contribution to recognizing the relevance of evolutionary biology and genetics for the study of social behavior—an area that had been dominated for decades by a purely environmentalist approach. I present here, however, my disagreement on certain matters. (His reply appears in the Perspectives *issue noted below.)*

One of our differences is over his belief that a sufficient knowledge of the human nervous system will lead to a true ethics. Perhaps my disagreement simply reflects thinking on a shorter time scale than Wilson about the future of neurobiology and sociobiology. But it still seems to me that the fundamental issue is one of logic, expressed in the naturalistic fallacy, which would exclude the possibility that science can ever provide anything approaching a complete solution to problems of values. At the same time, as elsewhere in this volume, I emphasize the conviction that science can contribute substantially, without being prescriptive, toward solutions to such problems.

I also present here my differences with some views expressed in Wilson's On Human Nature*: his belief in the virtually universal human need for religion, and in the capacity of evolution to provide a presiding myth for a new religion that will not conflict with science. This discussion leads to some brief thoughts of my own on the relation of genetics to free will and to epistemology. Finally, I chide Wilson for approaching human sociobiology almost exclusively in terms of universal characteristics of the species, while avoiding the controversial but highly relevant topic of human genetic diversity. Subsequent to the publication of this paper C. J. Lumsden and Wilson published a book on the coevolution of genes and*

Published as "The Importance of Human Diversity for Sociobiology" in Zygon 15 (1980): 275-293; reprinted in *Perspectives in Biology and Medicine* 26 (Autumn 1982):1 .

culture, in which they explore, quantitatively and in detail, the role of genetic diversity in contributing to cultural diversity (Genes, Mind, and Culture, *Harvard University Press, 1981).*

It was widely expected in the early years of genetics, at the beginning of this century, that this science would soon provide real help in solving many social problems.[1] The reaction to the subsequent disappointment, and to the underlying naive assumptions (e.g., "the gene" for intelligence), generated a widespread denial of the importance of genetic differences, and this extreme environmentalism has long prevailed in the social sciences and in liberal circles. After all, it was argued, we cannot define the role of genes with precision, and we cannot modify them (except by eugenic measures); so is it not more fruitful (and encouraging) to assume that most of our behavioral diversity arises from cultural influences, which we can modify? Indeed, this assumption is true for the large differences seen in the broad range of human cultures: but comparisons within a culture are another matter.

After this longer period of denial, Edward O. Wilson's *Sociobiology*[2] has now stimulated a broad renewal of interest in the role of genes in human affairs. Like Darwin's *Origin of Species*, this book synthesized a large accumulation of scientific information, and it defined a new field, of wide social as well as scientific interest. But unlike Darwin, who avoided involvement in the controversies over the revolutionary social implications of his theory, Wilson is the product of an age that has become very conscious of the impact of science on society, and he speculates in considerable detail about the future social implications of his field. Indeed, in a subsequent book, *On Human Nature*,[3] he presents these implications with some zeal, not simply as an inevitable by-product but as one reason for regarding sociobiology as a major discipline.

Wilson argues that if we wish to acquire a deep understanding of human social behavior we should not rely only on the observations and insights of the humanities and the social sciences; we must also look into our past evolutionary origins and our present genetic determinants. I agree. But at the same time, his comparative approach, as a naturalist, leaves out a large area: he concentrates on the universal characteristics of our species and virtually ignores our genetic diversity. Yet this diversity seems to me to be central for several of the issues that he considers. Let me first outline these briefly.

Wilson predicts that sociobiology will eventually provide a firm, objective foundation for ethics. I would suggest instead that the genetic diversity within our species increases our disagreements and conflicts, on ends as well as means, and because of these inevitable disagreements we will always require social negotiation in the development of our rules of conduct. Wilson goes on to emphasize that religions express a deep-seated, inescapable aspect of our biological heritage, and he struggles bravely to reconcile their aims with the scientific outlook. But in the end his projected solution—a new religion based on evolution as our presiding myth—seems to ascribe too much uniformity to our drive for religious belief. In fact, the wide diversity in our emotional patterns and needs has clearly made it possible for people to create, and to convince others of, widely varying attitudes toward religion (including its rejection). Finally, as the architect of a new field Wilson understandably concentrates on its future directions and benefits. But his predictions of benefits might be strengthened by focusing on one that is already available: even with our present limited knowledge of human genetics, our evolutionary origins carry the unavoidable implication that our species must possess wide genetic diversity in behavioral as well as in physical and biochemical traits; and recognition of this fact could help us to build our social policies on a more realistic base.

In this third point my greater attention to present benefits no doubt reflects my initial training in medicine, which has a more pragmatic outlook and a much shorter time-frame than evolutionary biology. But if sociobiology expects to help solve social problems, just as other branches of biology are being used to help solve problems of health, its aim is equally pragmatic. Let us begin, then, with Wilson's views on science and values.

Integrative Reductionism

On Human Nature offers us an exciting vision: When we understand the neurobiological basis of human motivation and action we will be able to fashion value systems based on reality rather than on illusions and false preconceptions. Wilson presents this proposition as a logical extension of integrative reductionism. This is the aspect of science in which our initial analysis of a phenomenon, formulated in terms of the obvious units, becomes much deeper when we can reduce it to analysis at a finer level of organization, i.e., to explanation in terms of the properties and interactions of the component elements within the original units. An example is study of the nervous system not only in terms of its integrated action, and the location of the groups of cells and connections responsible for those actions, but also in terms of the organization of individual cells and their component molecules. On the other hand, the reverse of this reductive process is not equally valid: the principles developed at the finer level of organization cannot by themselves predict the phenomena and the principles observed at the higher level.

This integrative approach is inherent in the scientific method. Indeed, it is responsible for the remarkable coherence of the resulting body of knowledge. Unfortunately the term reductionism has also been applied to an impoverished interpretation of scientific materialism that belittles the higher levels of organization—for example, the description of man as "nothing but" an animal, or a collection of cells, or a collection of chemicals.

In his discussion of integrative reductionism Wilson describes the natural sciences as a hierarchy in which each discipline is firmly based on its antidiscipline, that is, the next finer level of organization.[4] The social sciences, in contrast, lack roots in an antidiscipline. In his view they will become strong only when they develop such roots, and these obviously lie in sociobiology. While I agree that such integration will be valuable, I would question whether sociobiology is the sole antidiscipline to the complex social sciences, and whether the continuity between the two can ever become nearly as complete as the continuity between neighboring natural sciences. The problem is not simply the territorial resistance of the social scientists to an integration with biology; as I shall discuss in the next section, there are also more fundamental obstacles.

Limits to the Scientific Analysis of Values

The most important obstacle to a thorough fusion of the natural and the social sciences is epistemological. Many of the questions asked in the social sciences involve value judgments; and with such questions we cannot readily find an objective basis for identifying an answer as correct—as we can, in principle, with any questions about the nature of the material world. This distinction is well known in philosophy as the naturalistic fallacy—David Hume's principle that we cannot derive an "ought" from an "is."

Wilson explicitly questions this principle. In *On Human Nature* he proposes that the new knowledge will make it possible to fashion "a biology of ethics" and "a genetically accurate and hence completely fair code of ethics." Similarly, an understanding of the limbic system, and thus of the origin of our drives, will lead us to choose the "truer emotional guides" among various alternatives. These phrases seem to suggest the expectation that science will eventually prescribe an objectively correct ethics.

As Gerald Holton[5] has pointed out, here Wilson is joining the scientific tradition of Hermann Helmholtz, Ernst Haeckel, and Jacques Loeb:

that is, the assumption that the scientific method will be able to provide definitive solutions to the problems of society. However, they built on analogies and projections from early studies in organ physiology, and his base is much more sophisticated. Even with our present rudimentary knowledge it is clear that the levels of various hormones, and of the recently discovered neurohormones released within the brain, strikingly affect such features of our behavior as output of energy, sex drives, appetite, and mood. Moreover, future advances in neurobiology will surely provide increasingly fine-tuned knowledge of the material basis for those mental processes that give rise to value judgments. With such developments, in Wilson's view, the logical limitations imposed by the naturalistic fallacy will no longer be so self-evident or so absolute. I do not believe we can simply dismiss this view with the pejorative term scientism, since we will be considering a radically new kind of knowledge. But in trying to imagine how far neurobiology will be able to take us, I still see several reasons to doubt that it will be able to prescribe a correct ethics.

The main reason is that the criteria for what is good or right will not be deducible from the properties of any individual limbic system, or even from the shared, universal properties of human limbic systems: Interactions within the group play an indispensable role. Conceivably, detailed knowledge of an individual's limbic system (and of much more of his brain) could tell us why that individual attaches greater value to one rather than to another goal or activity, or why he balances immediate advantages against long-term advantages in a particular way. But this person's preferences will not be equally congenial to all other persons. Moreover, even individuals with identical goals will be in conflict when they compete for the same resource. Accordingly, the values that guide social behavior within a group will continue basically to be derived by a political process, whether of negotiated agreement or of imposed authority.

This is an aspect of ethics that Wilson seems to ignore. To be sure, the advance of neurobiology should eventually permit us to project more accurately the population distributions of alternative reactions to various possible restraints and incentives. But reducing in this way the error in the assumptions and predictions on which we build our social code is not the same as prescribing that code.

A second limit to the scientific analysis of values arises from the vast volume of data that would be involved in pursuing this analysis at the neurobiological level, for we can accumulate and process only a finite amount of knowledge, even with computers. And though one can conceive in principle of translating the votes of millions of persons into a neurobiological analysis of the mechanisms underlying each vote, converting the principle into practice is another matter.

To illustrate the problem, consider the widely quoted assertion that 1,000 monkeys typing randomly for 1,000 years would produce all of Shakespeare's works. This statement is in fact extraordinarily inaccurate. With random typing of the 26 letters of the English alphabet a length of only 15 units would have more different sequences (26^{15}) than the number of seconds in the history of the universe (taken as 15×10^9 years). Hence in all of time so far a monkey would have little probability of reaching even one predicted line of print! With this recognition of the difference between dealing with such numbers in principle and dealing with them in practice, we can hardly expect human decision-making to be defined in terms of measurements of activities of human brain cells.

We thus see two major limits to the biological analysis of ethical problems: the ultimately political origin of values, and the virtually infinite number of neural events involved. In addition, a third, probably minor limit is inherent in the predictive powers of science. We have known for over a century, since the development of statistical mechanics and the discovery of radioactivity, that the world is not strictly deterministic. Under its apparently deterministic macroscopic surface lie microscopic events that are predictable only statistically and not individually. Moreover, as the recently developed field of catastrophe theory has emphasized, in some systems a chain triggered by such a small event can have large, irreversible consequences for the system as a whole. In human populations such amplifications, with significant effects on history, could arise from various kinds of unpredictable events: the random occurrence of a particular mutation, the random fusion of two germ cells to produce a particular genotype in a future political leader, or the firing of a particular brain cell that swings a closely balanced decision.

I conclude, then, that in the cultural evolution of ethical systems the social interactions, and the processes within each individual, are too numerous and too incompletely deterministic to be adequately definable in neurobiological terms, even though they are all composed of neurobiological events. Hence the naturalistic fallacy would hold even for knowledge that penetrates to the limbic system. Accordingly, in this area we cannot expect a scientific process to replace the trial and error, and the compromise, of the political process—including the adaptations of religions to changing social circumstances.

Possible Positive Contributions of Science to Ethics

Having emphasized limits, let us now consider what positive contributions science can make to ethics. First, apart from the content of science, the method itself has an impact, by providing a powerful tool for making our predictions and our assumptions more reliable (i.e., more concordant with reality). An individual choosing between alternative actions not only weighs their immediate values but also often has to consider (though not always explicitly) their possible, more remote consequences. The same is true of group decisions: in the development of general value systems cultures build on predictions about the consequences of alternative kinds of actions, as well as on assumptions about human nature; and within this framework every public policy implies a prediction of consequences. The success of science, with its emphasis on testing predictions and assumptions against reality, has clearly fostered a more pragmatic, situation-oriented approach to ethics in the modern world. But in making this contribution science can be only an adjuvant to, rather than a replacement for, a political process (using this term in a broad sense).

In addition to this general role of the scientific method in helping us to build on reality, the content of sociobiology may also have a more specific role: that of helping to enculturate moral motivation by convincingly legitimizing, within a materialist and evolutionary framework, the principle of a moral consensus. As Ralph Burhoe has emphasized,[6] the discovery of our origin by natural selection, rather than by purposeful divine creation, has contributed to a weakening of the moral consensus in the modern world. This discovery not only destroyed a traditional foundation of morality, without providing a satisfactory substitute; it also seems to many persons to have entailed as its logical consequence an extreme moral relativism. But this view is based on the belief that the fundamental law of evolution is the unrestrained competition of "Nature red in tooth and claw." And we now recognize that this is a misconception: Evolutionary mechanisms yield altruistic as well as selfish drives, and both are essential parts of our nature as a social species. Indeed, the recognition and analysis of these mechanisms is one of the main contributions and aims of sociobiology.

Nevertheless, recognizing a biological basis for altruism is still a long way from providing a foundation for a moral consensus. The only firm mechanism, that for "hard-core" altruism, applies only to kin with shared genes. "Soft-core," reciprocal altruism, with a much broader range of beneficiaries, is also recognized in sociobiology, but its relation to the general problem of ethics is far from clear. Many scholars have tried to derive ethics from evolution, ever since Herbert Spencer and Thomas Huxley, but their success has not been impressive.

I would suggest that we might be more successful if we set our sights lower, that is, if we try to apply an evolutionary perspective in a conditional mode, comparing the long-term consequences of alternative attitudes, rather than trying to apply it in an imperative mode, specifying detailed personal obligations. Sociobiology cannot specify any particular degree of altruism as correct, and it cannot even establish species survival as a cardinal value; but it can predict that if we are to survive and function as a social species we must agree on a set of ethical standards. Within this framework any particular standards would continue to be evolved by a cultural, political process—and in this process surely few would question the desirability of having our species survive, even though the "correctness" of this goal cannot be rigorously proved. In other words, sociobiology can say that the idea of right and wrong, with its implications of socially sanctioned obligations and restrictions, is not simply an artifical cultural construct, imposed by those in power. It is rooted in genetically conditioned drives, shared by all people, though varying widely in intensity from one person to another. The genes thus provide the potentiality, and the need, for moral behavior.

This principle does not seem very novel, nor does it offer much help in the eternal human problem of choosing specific values. But it offers us a philosophic basis for developing those values within the framework of a respected social order, rather than within the socially destructive framework of extreme moral relativism. Moreover, this approach builds on postulates that are thoroughly consistent with the scientific world view, without the need to invoke the transcendent.

While I thus conclude that sociobiology can have only an adjuvant role (but a valuable one) in the development of ethics, Wilson seems to expect a larger role. And one could argue that even if his claims are too optimistic the interest that they stimulate may be useful. On the other hand, there is a danger. For example, a few decades ago some pioneers in molecular genetics were tempted to speculate proudly about the future miracles of genetic engineering—but as gene manipulation drew closer these fantasies bounced back painfully. To avoid excessive hopes, and anxieties, perhaps it would be best to let the science of sociobiology advance without too much effort to anticipate its social applications. For we have limited capacity to predict future developments in science, and even less to predict their social consequences.

Implications of Genetically Encoded Information for Epistemology

Modern biology has further implications for ethics, derived from its relevance for our understanding of epistemology. This relevance arises from the emergence of a fundamental new concept in molecular biology: the storage of information within molecules. This concept has revealed the continuity, at a molecular level, between several kinds of information. Thus genotypic (inherited) information is stored as a program in the sequences of DNA; this program is expressed (epigenetics) during embryonic development, including that of the brain; and acquired (learned) information is stored in the form of molecular modifications, as yet little understood, in the distribution and the functional properties of the synaptic connections in the neural network of the brain. We thus have a coherent framework for understanding the process by which interactions of genes and environment create the phenotype.

Indeed, a particular activity, such as the formation of a specific enzyme in a bacterium or the use of a specific call by a bird, may be rigidly determined by genes in one species while in another the response may involve learning as well. With phylogenetic ascent the ratio of learned information to inherited information increases: that is, the genes increasingly provide ranges of behavioral potential rather than specific behavior. At the extreme the human species can process information in a uniquely subtle and complex way, as a result of selection for adaptability to varying circumstances more than for adaptation to specific circumstances. Nevertheless the "hard-wired" information coded by the genes still has a role, hidden beneath our learning.

With this recognition, that we possess genetic information and that it merges with our acquired information about the external world, it is now clear that Kant's epistemology, involving a priori, inborn categories of knowledge, was much closer to reality than British empiricism. The evolutionary survival of our species has depended on the ability of individuals to interact effectively with a challenging environment, and these interactions could not be effective unless our genes programmed our nervous system with the necessary internal information on which to graft our learned information. We also could not survive unless the resulting information about the external world was reasonably reliable (though not necessarily infallible). Hence as infants make contact with their surroundings they develop the concepts of space, time, and causality that are necessary for effective interaction. They also correlate the evidence provided by their five senses. These aspects of growth and development must involve appropriate prewiring in the brain, as well as subsequent modification by experience.

A similar evolutionary principle obviously applies in linguistics, where it has been amplified into a major thesis. We are not born to know a particular language, but our intense selection for improved communication has evolved hard-wired connections in the brain that create the capacity for a complex, rich language.

In the same way, our functioning as a social species requires that our brain contain a prewired general foundation for ethical judgments. The details then emerge as a social construct—developed in response to our needs, based on our biological natures and our cultural histories, changing as part of cultural evolution, and not dependent on any immanent purpose in the universe. Sociobiology thus provides a naturalistic explanation, in terms of gene-environment interactions, for the origin of ethics. In this perspective ethics is partly deontological (but with a genetic rather than an extramaterial source), and at the same time it is partly utilitarian (i.e., calculated in response to environmental opportunities and constraints).

This perspective provides only a soft foundation for ethical systems. It therefore may not satisfy those philosophers who seek something rigorous and sharply defined, such as Kant's categorical imperative or Rawls's postulates. But at the risk of ignoring an enormous and sophisticated literature I would suggest that ethics, as the product of biological and cultural evolution, does not lend itself to rigorous philosophical argument, and it may benefit from the naive approach of biology. Because of the immense behavioral plasticity that has evolved biologically in our species we can adapt our conduct, in cultural evolution, to a broader range of circumstances than any other species. We can therefore experiment with a far broader range of social patterns. Our evolutionary success depends on this flexibility: on balance, compromise, and continual adaptation to changing specific circumstances, rather than on uniformity and consistency. And the resulting patterns are all built on the range of genetically determined potentialities, within the population, that has been programmed by natural selection to be adaptive for our survival and for that of our progeny. Wilson makes this point bluntly: "Morality has no other demonstrable ultimate function [than to keep] the human genetic . . . material intact" [3, p. 167]. Dawkins has developed this point in detail in *The Selfish Gene.*[7] But it does not follow that morality is "nothing but" genetics!

Sociobiology and Utopias

I would now like to turn briefly to the implications of sociobiology for a special set of moral problems: those recently created by our development of a complex technology. We have belatedly recognized that technology has costs as well as benefits. In response, an alienated counterculture has revived the romantic notion that man was free until society fettered him with unnatural bonds. This approach leads to a retreat from reason as well as from reality. Another alienated group are the neo-Lysenkoists, opposed to all applications of genetics to human behavior. For them sociobiology is a reactionary force that discourages social change and even supports racism. Yet as I read Wilson I do not find him concerned with defending any particular political or economic system; I find him concerned with learning how to build, whatever the system we choose, on a deeper understanding of human nature.

Nevertheless, it is not hard to find a realistic reason for the opposition from political ideologues: Their utopias are built on assumptions of human malleability and perfectibility, and a sociobiological perspective does threaten these assumptions. Like the insights of Freud, and of the masters of literature and the great religious leaders, the sociobiological approach recognizes that tensions and conflicts are an unavoidable price of our evolutionary gifts of social interdependence, behavioral plasticity, and diversity. Hence society, regardless of its structure, will always be struggling to promote a balance between our aggressive and our altruistic drives. Moreover, as Wilson notes, aggressiveness has many forms, and some are essential for the creativity and the dynamism that have built up civilization.

But while Wilson recognizes the conflict and tragedy inherent in the human condition he does not emphasize it; he prefers to focus optimistically on the future contributions of sociobiology. As Charles Frankel has pointed out,[8] this perspective is ironic. In presenting sociobiology not only as an area of scientific investigation but as the path to a true system of ethics Wilson resembles his severest critics in himself having a utopian vision, though with a biological rather than a political base.

Human Diversity

Thus far I have been raising largely philosophic and social issues, concerned with the validity of various extrapolations from sociobiology to human social problems. Now I would like to consider an aspect of the scientific content of sociobiology as presented by Wilson, whose almost entirely ethological approach concentrates on the universal behavioral characteristics of each species (or larger taxonomic group); he pays little attention to behavioral diversity. This approach seems to me to result in an imbalance, with several important consequences.

I have already discussed one consequence of Wilson's focus on universals: an underestimation of the role of political negotiation in creating ethical rules. A more important consequence is neglect of valuable present contributions that would follow from recognition of wide genetic diversity in behavioral traits. These inborn differences—in intellectual capacities, motor skills, special talents, drives, preferences, and emotional responses—are obviously relevant to our society's approach to many urgent problems of distributive justice, including education, job allocation, and economic rewards. Indeed, the most valuable ultimate contribution of biology to the social sciences may be to identify more precisely the genetic and the environmental factors that contribute to our differences.

The results will not prescribe how society should handle our biological diversity. But they can improve our ability to maximize individual self-fulfillment, for the better we understand inborn differences the better we can fit the environment to the genotype. As Theodosius Dobzhansky[9] emphasizes, we jeopardize the quest for greater social equality if we rest it on the assumption of biological identity, rather than on the foundation of moral and political principles: for the former, but not the latter, is vulnerable to empirical disproof. We can legislate our social institutions, but not our genes.

Nevertheless, because genetic diversity sets limits to equality of achievement, it is widely regarded today as negligible, or else as an unfortunate cost of evolution—like painful childbirth as the price of a large brain, or susceptibility to back strain as the price of bipedal posture. Biology can help us to recover a realistic and sensible attitude on this matter, for it is axiomatic that genetic diversity has great value for species survival. In addition, diversity is indispensable for the development of a rich and interesting culture. What an incredibly dull world it would be if we were all genetically identical!

This aspect of sociobiology seems to me central. Wilson, in contrast, has concluded that altruism is the central theoretical problem of sociobiology. And at the moment the latter topic looms large because a reasonable and testable theory has recently been developed. But in the long run diversity seems to offer more extended horizons, at least for human sociobiology.

Human diversity, of course, has become the subject of intense political controversy. And since Wilson was already offering a challenge to other widely held beliefs, it is understandable that he would not wish to look for additional trouble. But he goes beyond merely sidestepping the issue when he states that genes have only a "moderate" influence on mental ability [3, p. 198]. Our present knowledge does not warrant such a definite conclusion. In fact it supports a high probability of a very substantial influence of genes—at least 50 percent of the observed variance within the populations tested. Similarly Wilson notes that reassortment of genes will permit ordinary parents to yield an exceptionally talented genotype [3, p. 198]. This is technically correct; but the statement might mislead a reader since it seems to imply that abilities are randomized from one generation to the next. In fact, they are not: Even though recombination of genes allows individual progeny to deviate broadly from their parents, the average genotypic level of the progeny in a family will ordinarily be close to the midpoint of the two parental values. Genetic diversity is thus clearly a significant factor in human social behavior, and it would be unfortunate if political sensitivity should inhibit its inclusion in sociobiology. Indeed, since the component elements of human social phenomena include individual patterns of behavior, as well as the universals of our species, sociobiology can hardly claim to be the antidiscipline of the social sciences unless it takes account of both.

In two other areas that Wilson discusses, religion and free will, diversity (in emotional patterns and needs) is also pertinent, as we shall note below. And as a final cost, failure to face squarely the topic of genetic diversity deprives Wilson of the opportunity to rebut the greatest source of resistance to his thesis: the fear that any attention to genetic differences might distract attention from, or might even undermine, the goal of eradicating inequitable social practices.

The historical cause of this fear is quite understandable. The tragic consequences of earlier applications of evolutionary and genetic principles to society by Social Darwinists, eugenicists, and racists justify concern and vigilance. But if we examine this history more closely we will find that these destructive early applications were based on premature extrapolations, or on gross distortions that should be recognized as pseudoscience rather than as science. A careful, stepwise accumulation of knowledge of sociobiology should protect us from such distortions in the future, rather than promote them.

Free Will and Determinism

Let us now turn to Wilson's discussion of a central paradox: free will and determinism [3, p. 77]. He suggests that we appear to have free will simply because the human mind is so complex, and our social relations so intricate and variable, that detailed individual histories cannot be predicted in advance; yet the paradox of freedom and determinism is resolvable in theory and might be reduced to an empirical problem. I agree, but I would suggest that these propositions are incomplete. The source of apparent free will is not simply the complexity of the human mind; it is also the genetic diversity of human minds. If we were all genetically identical we would behave very similarly—indeed, much more similarly than identical twins do today, for they are exposed to the diverse models of the behavior of many other genotypes. And the more predictable our individual behavior, the less free will, as we now understand it, would remain.

Accordingly, while theological formulations of the problem of free will and determinism have led to postulates of an autonomous soul, able to choose between virtue and temptation, if we wish to consider the problem in biological terms we must see free will as an expression of the complex interactions between diverse genotypes and diverse environments. The basic question should then be framed quite differently: not how much of our action is free and how much is determined, but (1) how much my reactions and my choices in responding to competing stimuli differ from those of my neighbor, (2) how much of a change in the strength of these stimuli is needed to eliminate such a difference in responses, and (3) how much each difference in our patterns of response is due to differences in genes and how much to past exposure to different environments.

The element of freedom in our behavior can thus be divided, like all phenotypic traits, into genetic and epigenetic components. There may be a third source of variation, "developmental noise"—a phenomenon readily seen in our physical phenotypes as the persistent effect of random molecular fluctuations on a developing organ (such as differences in the fingerprints of identical twins). It seems reasonable to suppose that in the function of the nervous system a parallel kind of noise—the unpredictable, chance firing of a critical neuron—occasionally also affects actions, thus contributing (probably very slightly) to what appears to be free will.

Sociobiology and Religion

The traditional theological problem of free will and determinism brings us to the final chapters in *On Human Nature*, which consider science and religion. I find it hard to comment on these chapters, for, as is frequent in this perennial controversy, the term "religion" and the associated beliefs do not have clear or consistent meanings. Durkheim's definition, "consecration of the group" [3, p. 169], would not have to be stretched very far to include the fans (derived from the word "fanatic") wildly cheering the local basketball team, or the staid members of the National Academy of Sciences politely applauding this year's recipient of the U.S. Steel Award in Molecular Biology. Moreover, Wilson accepts an anthropologist's estimate that mankind has produced on the order of 100,000 religions. On the other hand, elsewhere he states that his concern is "real" religions—presumably ranging in our culture from fundamentalist orthodoxies to Ethical Culture. The protean nature of religious belief (and also Wilson's tact) thus makes the discussion less tightly organized and reasoned than his earlier discussion of ethics.

Nevertheless, Wilson has introduced a novel and interesting approach. Instead of either defending the value of faith or criticizing its conflict with evidence, he focuses on the religious impulse as a product of human evolution. He concludes that religions serve an inescapable set of emotional needs, determined by our genes. These services include mechanisms for encouraging altruism and promoting adherence to the group's moral norms, and for providing several sources of inner security: a sense of individual and group identification, submission to hierarchical leadership, comfort in time of distress, confidence in time of battle, a sense of purpose and destiny, a promise of future salvation that removes the dread of death and makes present suffering more tolerable, and a magic influence over external events. In addition, symbolism, myth, and ritual are used not only to lend affective support to these beliefs but also to cultivate esthetic sensibility and sensitivity to human feelings, in ways that are missing from the cool, rational approach of science. In Wilson's view a detailed understanding of the biological basis for these many emotional needs will permit us to develop a new kind of religion, one that will eliminate the traditional conflict with science.

I would suggest that this discussion might have been more sharply focused if Wilson had not treated the body of religion as a whole but had separated its function of enculturating moral values from its several other functions. Even the most intransigent atheist would agree that the need for a moral consensus is universal. The various other services of religion, in contrast, meet needs that are less than universal, and in ways that often conflict with science. Hence many liberal theologians now concentrate on preserving what they see as the heart of the religious tradition: the culturally evolved wisdom about man's relation to man.

The problem of finding a reconciling format is illustrated by the persistence of prayer in religious services. It is understandable that this traditional ritual, through its influence on feelings and attitudes, continues to give satisfaction to the supplicants, even if they no longer expect it to influence external events. Yet many scientific materialists, fearful of any concessions to irrational forces in our society, are made uneasy by the ambiguity between symbolism and literal content in the words of prayer: because it encourages a lingering hope of a magical influence at times of desperation, it undermines efforts to build policies on reality.

Recognizing the problem engendered by this split between two concerned groups, and considering religion inevitable (for biological reasons) as a major social force, Wilson tries sympathetically to seek compromise and reconciliation. Indeed, it seems to me that he even exaggerates the role of organized religion in this country today. Emphasizing the large proportion of professed adherents in the population, he fails to differentiate between intellectual leaders and followers, and between real commitment and social convenience. Even more, he underestimates the roles of other institutions (the family, education, law) when he describes religion as "above all the process by which individuals are persuaded to subordinate their immediate self-interests to the interests of the group" [3, p. 10]. Yet beneath his conscientious effort one cannot help recognizing some ambivalence. He speaks at one point of the large fraction of the population that adheres to a traditional faith, yet at another of the "fatal deterioration" of the traditional myths. Similarly he dismisses as obscurantist the search of Theodore Roszak for meaning in the "dark, shadowy tones of religious experience" [3, p. 10], but in a later chapter he is much more sympathetic. In the end, by a circuitous route, he winds up with a classical replacement of theology by science. The coup de grace comes from sociobiology's capacity to explain the evolution of the religious impulse as a wholly material phenomenon [3, p. 192]. But it is not clear why this kind of evidence from science, about the origins of religions, should threaten them any more than earlier evidence conflicting with their content.

On the other hand, the conclusion that our need for religion has inescapable biological roots leads Wilson to the hope that we can finally reach a reconciliation by a new approach that makes evolution the "presiding myth." He thereby seems to be seeking an essentially single modified religion, as logically coherent, as consistent with reality, and as universal as the scientific world view on which he builds. But, as I noted above, this expectation suffers from neglect of our diversity. As with all behavioral traits, the genetic factors that contribute to our reactions to religion will vary widely. Some individuals are more discomforted by uncertainty and by lack of answers to "ultimate" questions than by the inconsistencies that arise when traditional religions provide the answers; with others the opposite is true. Moreover, people obviously vary enormously in their receptivity to various kinds of reasons for accepting a belief: whether because the evidence is convincing, or because most of their neighbors share the belief, or because it makes them feel better, or because they think it will encourage them and others to act better. Finally, the cultural milieu strongly influences individual "freedom" of religious choice: the term "parochialism" reflects the widespread assumption that the religion of one's parents is obviously the right one.

It is clear, then, that recognition of the genetic basis of our emotional needs does not tell us how many people will continue to try to meet these needs in terms of theistic religions, and how many will prefer to extract from science a more coherent, but also more austere, world view, having little continuity with these traditions. Perhaps the changing attitudes toward aggression imposed by civilization can provide a helpful analogy. Thus because of its genetic basis, aggression will always be with us; but as societies try to persuade people to alter their patterns of expressing aggression from those of our neolithic ancestors some individuals will always prove to be much less responsive than others.

Given our genetic and cultural diversity, it is hardly surprising that individuals and groups have met their religious needs in many ways. Stoicism, for example, was a secular religion without the postulate, so prominent in the Judeo-Christian tradition, of a transcendent creator; and the major Eastern religions also place little emphasis on a conscious god watching over us. And despite advances in sociobiology, the future of religion seems likely to continue to be pluralistic rather than monolithic, involving different patterns that meet different individual needs, rather than a single pattern that achieves thorough consistency with science. Many people will continue to postulate a source of purpose and identity that transcends the material world. But for others this concept is too hard to reconcile with man's emergence as a chance product of evolution. The latter group may be no less concerned with the need to transcend immediate, hedonistic self-interest, by dedication to some goal outside oneself.

However, if we wish to promote clear communication it would be a dubious solution to try to soften the conflict by subsuming both the religious and the scientific approach under the term "transcendence," redefined in the broader sense of self-transcendence. In traditional religious usage the word refers to transcending the material world, and it is thus a euphemism for supernatural. The fundamental problem of reconciling religion and science will not be solved by evading this issue.

I also do not share Wilson's confidence that sociobiology can achieve this reconciliation by leading to an evolution-based religion. On the contrary, the contribution of sociobiology to moral values might be weakened if we further link the field to the other aspects of religion—matters to which science can contribute very little. More broadly, treatment of the evolutionary epic as a myth, as he suggests, might weaken science without strengthening religion.

Wilson resembles Freud in seeking to analyze rationally the basis of nonrational behavior, but he works on a different level. He concentrates on the evolution of indoctrinability and religious faith, as well as of altruism. But it is not evident how sociobiological evidence on the origins of the religious impulse will help to solve the problems that religions now wrestle with, or to lessen the conflict between faith and reason. In contrast, Freud, working at the level of observed behavior and with a physician's concern with the present, uses ontogeny rather than phylogeny to explain the irrational elements in our behavior. He sees religion as the expression of a persistent, unconscious infantile yearning for dependence on a protective, powerful parent. And since his general aim is to replace childhood fantasies with reality-based adult behavior, he is not very sympathetic with the religious tradition. Freud is thus more tough-minded than Wilson, both in recognizing a deep conflict between the perspective of science and that of traditional religions, and in recognizing the implications of the inherence of tragedy and conflict in human nature.

Conclusions: The Scope and the Limits of Sociobiology

Wilson has convincingly established the evolutionary biology of social behavior as a major field, and he has thereby done much to stimulate interest in the role of genes in human behavior. However, in his focus on the universals that characterize each species, I believe he has neglected individual genetic diversity. This diversity is especially important in our species, where it must strongly influence the paths of cultural evolution.

This neglect has a serious effect on Wilson's discussion of ethics. He suggests that when we can delve in detail into the aspects of the human brain that are concerned with motivation Hume's sharp distinction between "is" and "ought" will no longer be valid, and we will be able to develop a completely fair system of ethics. I question this conclusion. Though science can help us to evolve better rules and to make better individual choices by improving our predictions of the consequences of alternative actions, it seems very doubtful that we will ever be able adequately to specify in neurobiological and genetic terms the elements that enter into an individual's value judgments. Moreover, given the heterogeneity of our population, it is even more difficult to see how knowledge of the average limbic system, however detailed, could displace a broadly political process in forming rules of conduct. Indeed, the greatest accomplishment of applied sociobiology may be almost the opposite of prescribing ethics. Instead, by recognizing the importance of genetic differences, and the inevitability of genetically based conflicts within individuals and between individuals, sociobiology could supply a corrective to the illusion that progress in science and technology, or in politics, can lead to a completely harmonious society based on the moral perfection of man.

Individuality is pertinent also to the discussion of religion. Emphasizing that deep, genetically based emotional needs underlie religion, Wilson hopes that evolution will become the presiding myth of a religion that will meet these needs without conflict with the scientific world view. I would suggest, however, that attention to human diversity would favor a more pluralistic solution. Moreover, in avoiding discussion of the political aspects of social behavior, Wilson fails to note how much politics (in the usual sense) has displaced religion and ethical analysis, in recent centuries, as a source of our rules of conduct. Finally, in emphasizing the future applications of sociobiology to man, he virtually ignores implications of our present knowledge—especially of diversity.

Alfred North Whitehead has described philosophy as the critic of our abstractions. In the area of morality and social policy biology will undoubtedly play a parallel role, as a critic of our assumptions. Today sociobiology, focusing on evolutionary origins and dynamics, provides the key. But insight into our origins offers us much less guidance than knowing how we function. We can therefore expect the future contributions to come increasingly from neurobiology, linked to sociobiology by the still nascent field of neurogenetics.

It is impossible to foresee how far sociobiology and neurobiology will go in improving our ethical systems and in promoting their acceptance. But we must recognize limits. Biology can provide firm facts and can reveal underlying mechanisms, but these are only a foundation. Not only for those who feel a need to invoke the transcendent, but equally for those who do not, the biological description of human nature can only be coarse-grained: Analysis of gene-environment interactions is no substitute for such concepts as poetry, inspiration, and love.

In the search for the biological roots of human behavior what is justly feared is that an integrative reductionism, intended to broaden our perspective, could slip into the kind of reductionism that would narrow that perspective. To avoid that pitfall sociobiology, like the humanities, surely must focus on our individuality, as well as on our common humanity.

Notes and References

1. M. Bressler, "Sociology, Biology, and Ideology," in *Genetics*, edited by D. C. Glass (New York: Rockefeller University Press, 1968).
2. E. O. Wilson, *Sociobiology: The New Synthesis* (Cambridge, Mass.: Harvard University Press, 1975).
3. E. O. Wilson, *On Human Nature* (Cambridge, Mass.: Harvard University Press, 1978).
4. E. O. Wilson, "Biology and the Social Sciences," *Daedalus* 106 (Fall 1977): 127–140.
5. G. Holton, "Sociobiology: the New Synthesis?" *Newsletter on Science Technology, and Human Values* 3(October 1977): 28–43.
6. R. W. Burhoe, Editorial. *Zygon* 13 (1978): 250–256.
7. R. Dawkins, *The Selfish Gene* (New York: Oxford University Press, 1979).
8. C. Frankel, "Sociobiology and Its Critics," *Commentary* 68 (July 1979): 39–47.
9. T. Dobzhansky, *Genetic Diversity and Human Equality* (New York: Basic Books, 1973).

10

New Foundations for Evolution

Several fields in biology that emerged after *Darwin have provided powerful and novel kinds of evidence for his theory. This paper proposes that in the defense of science against the so-called creation scientists this material may be more convincing than the gap-filled paleontological record.*

In the first part of the paper I describe how classical genetics, filling a major hole in Darwin's argument, led to the much more widely accepted neo-Darwinian synthesis. The further contributions of comparative biochemistry and of microbiology, important but less well known, follow. I then concentrate on the dramatic impact of molecular genetics. This field has provided an entirely unexpected kind of evidence for the genetic continuity of the whole living world, far more direct than any previously available. It has also developed a "molecular clock" for better estimating the sequences and timing of the branches in phylogeny. Finally, its analysis of evolutionary mechanisms goes beyond kinetics and probes into concrete molecular processes. In this way evolutionary biology, which has unified the many branches of experimental biology, becomes itself unified with them.

In the century since Darwin developed his theory, largely on the basis of comparative morphology and paleontology, genetics and comparative biochemistry have provided a great deal of further support. And in a dramatic further advance molecular genetics has now yielded a new, more direct kind of evidence for evolutionary continuity, extending from bacteria to man. Indeed, unless we assumed that continuity the study of molecular genetics in bacteria would not help us to understand human cells.

Published as "Molecular Genetics and the Foundations of Evolution" in *Perspectives in Biology and Medicine* 28 (Winter 1985):251–268.

Yet various groups remain skeptical, for various reasons. Religious fundamentalists in the Judeo-Christian tradition object that the evolutionary view of man's origin destroys an indispensable basis for morality. Extreme egalitarians have difficulty with the implication that the genetic diversity within each species, on which natural selection depends, must include mental (as well as physical and biochemical) traits in man. Some literary people, following the line of Arthur Koestler, falsely ascribe to science the goal of discovering absolute truths, and they then criticize evolution for failing to meet that goal.[1] And while a distinguished philosopher of science, Karl Popper, accepted evolution as a fact, he questioned whether Darwin's theory (even in its modern, neo-Darwinian version) meets the criterion that he has proposed for distinguishing a scientific from a metaphysical theory: the ability to generate falsifiable (i.e., testable, refutable) predictions.[2] Popper has now conceded his criterion was too rigid,[3,4] but unfortunately, creationists have continued to draw support from his original position.[5]

A recent poll of a representative sample of Americans[6] illustrates the extent of resistance to the theory of evolution: 44 percent of the respondents believed in the special creation of man occurring within the past 10,000 years, two other groups conceded a longer time scale or else accepted the theory of a *directed* evolutionary process, and only 9 percent accepted the scientific conclusion that our species has evolved by undirected natural selection. Though this result is discouraging it is not hard to understand. Scientific ideas on man's origin are relatively recent, while religious ideas carry the weight of long tradition, have much more emotional appeal, and offer a simpler basis for a moral consensus. No wonder so many people find these ancient, poetic myths about man's origin more credible and more satisfying.

Nevertheless, since the question of the origin of our species is a question of biology, only objective scientific inquiry, divorced from moral preferences, can provide an answer that corresponds to reality. And since nature has the last word on such questions it is hard to doubt that the scientific answer will ultimately prevail. But "ultimately" may be a long way off; for although liberal religion is primarily concerned with questions of values, and has given over to science its earlier function of also trying to explain the world of nature, that is not true of all religions. Meanwhile, the tensions between science and myths are likely to become worse, as advances in genetics, neurobiology, and sociobiology further contradict treasured preconceptions—political as well as religious—about human nature.

Evolution is thus central to our attitude toward reality and to our assumptions about human nature. It is therefore essential, for the future harmony of our society, to try to teach the subject more effectively. In most of its development evolutionary biology has depended on morphological homologies, both in the fossil record and among living species; but this approach has not revealed the continuum of transition forms between species that Darwin predicted. Moreover, while he expected further research in paleontology to fill in the gaps, we no longer entertain that hope. But now, at last, molecular genetics has provided a direct, radically different kind of evidence for such continuity.

So far, however, this powerful evidence has penetrated very little into the introductory teaching of evolution and into debates with the creationists. In the recent spectacular legal victory of the American Civil Liberties Union against a creationist law in Arkansas, and also in recent books, the defenders of evolution have continued to focus almost entirely on the geological time scale and on the paleontological record, as in the Scopes Trial in 1923.[7] It is surely time for our teaching to balance this approach, without decreasing our appreciation for Darwin's remarkable achievement. Not only does molecular genetics provide the most convincing evidence for evolutionary continuity, but this evidence should interest a public that is well aware of the power of this science in other areas. I will therefore review some of the contributions of molecular genetics—as well as those of classical genetics, comparative biochemistry, and microbiology—to evolutionary biology.

Darwin's Problem with Variation

Darwin's theory has two major components: variation and natural selection. The latter received virtually all the attention, because of its courageous philosophical and religious implications. But the basic theme, dramatized as "survival of the fittest," has been accused of being a mere tautology—and it would be, if it were concerned only with the obvious idea of differential individual survival. Darwin's great accomplishment was to link that idea to heredity, thus creating the much more consequential idea of net differential reproduction.

The really radical component of the theory, then, was the assumption of endless herditary innovation, on which selection could act. But this was an ad hoc assumption. Wrestling all his life with this problem, Darwin came up with a mixed view: "hard" inherited variation, arising without direction by the environment, seemed likely to be the main source of novelty; but everyday observations seemed to point also to "soft" inheritance, responsive to use and disuse.[8] He therefore developed a logical but useless theory of "gemmules"—particles of inheritance in body cells that were responsive to use and disuse, and that were released to the germ cells and thus able to influence the next generation.

To be sure, Darwin did lean heavily on artificial selection, which obviously involves hereditary variation. But that process has a serious weakness as a model for natural selection. We now know that artificial selection depends largely (and it might conceivably have depended entirely) on genetic recombinations, arising from the range of genetic variation already existing within a species. Evolution of the enormously diverse living world from a common ancestor would require a much more thoroughgoing kind of genetic innovation. And since Darwin wrote *The Origin of Species* before the emergence of genetics, he had no evidence, or even a plausible mechanism, for explaining such innovation.

Darwin therefore could not proceed within the usual framework of science—that is, by means of a stepwise series of hypotheses, predictions, and confirmations. Instead, he had to make a large conceptual leap. His theory was thus in a sense premature. On the other hand, since some philosophers criticize the theory for its lack of testable predictions, one might also say that it made a grand prediction: a hereditary process that would reconcile the paradox of breeding true and yet creating novelty.

To appreciate how the development of genetics solved this problem, let us engage in a fantasy and pursue a hypothetical rearrangement of history, imagining that no one dared to propose the theory of Darwin and Wallace until it had a testable foundation in genetics. The two fields would then have arisen in a logical order.

A Hypothetical Scenario

The first step toward filling Darwin's big gap was DeVries's discovery of mutations, in 1900. Hereditary variation could then be seen to arise in two different ways: mutation provides the ultimate source of novelty, and the reassortment of genes in sexual reproduction enormously amplifies this variation. Nevertheless, for decades the actively growing field of genetics had little to contribute to evolution. Geneticists believed that species arose by giant mutations ("saltations"), rather than by the gradual changes invoked by Darwin, whereas evolutionists considered mutations to be exceptional monstrosities and assumed that the gradual steps of evolution must have arisen by some other mechanism.

Genetics and evolutionary biology were finally linked in the late 1920s, when the "particulate," Mendelian mode of inheritance, based on discontinuous traits, was shown to apply also to traits exhibiting continuous variation. Their apparent continuity arises from polygenic inheritance, in which many genes and their interactions with the environment contribute to the final numerical value of the trait. At the same time, Fisher, Haldane, Chetverikov, and Wright developed the quantitative science of population genetics.[9] This discipline provided a powerful new approach to evolution: measurement of the factors that cause these frequencies to change.

One of these factors is genetic drift: a process in which the chance geographic separation of a small population limits its gene pool. As a result the isolated population retains some mutations that would have been diluted by competing alleles in the major population, and these nonadaptive mutations influence the directions of further evolution toward valuable, adaptive traits. Recognition of genetic drift as a major process has greatly broadened the scope of evolutionary biology. Thus, although evolution as a whole depends on the selection of adaptive properties, not every step must be adaptive. This concept helps dispel the mystery of the survival of the intermediate steps in the evolution of a complex organ, such as the eye. Moreover, since genetic drift permits survival of individuals that might not have been the fittest in the original population, it removes evolution even farther from the tautology of "survival of the fittest."

Population genetics soon become the key discipline in evolution, and it changed profoundly our understanding of the nature of biological populations. Earlier biologists, under the influence of Aristotelian essentialism, had long characterized species (and races) in terms of an ideal type or essence, ignoring the presumably trivial individual deviations from the type. But the concept of uniform entities, though essential in physics and chemistry, is not appropriate for describing natural biological populations. In particular, genetic studies of biochemical traits (e.g., allelic forms of an enzyme) showed that a large fraction of genes exhibit polymorphism (multiple forms within a species). Each species thus conserves much more genetic diversity than meets the eye, and this diversity is now seen as a crucial rather than a trivial feature. As Mayr[10] has emphasized, this shift from a typological to a populational view has been one of the most important conceptual advances in biology.

Microevolution and Macroevolution

The development of population genetics revealed the true nature of geographic races, within both animal and plant species: these are populations whose prolonged reproductive separation has led to accumulation of significant statistical differences in their gene pools. Extension of this divergence would eventually create separate species by giving rise to reproductive incompatibility (because of incompatibility in behavior, genital fit, chromosomal organization, or perhaps histocompatibility antigens).[10,11]

Population genetics also made possible the direct demonstration of microevolution, that is, evolution within a species (or occasionally yielding a closely related species), and in contemporary time. For example, in localities where industrialization darkened the tree trunks an originally light-colored species of moth became predominantly dark. Whereas such shifts had earlier been ascribed to physiological adaptation, the mechanism was now found to be genetic adaptation (i.e., selection for an initially rare genotype).

These developments converted the process of natural selection from a bold hypothesis into a mere description of fact, just as the discovery of capillaries did for Harvey's bold hypothesis of the circulation of the blood. The extrapolation to macroevolution, which creates the enormous diversity of the living world, was logically compelling, and it was supported by the more or less closely graded morphological homologies in the phylogenetic trees. Essentially all biologists have accepted the result of this synthesis of evolution with genetics—the so-called neo-Darwinian model of evolution—as the central organizing principle in their field.[12,13]

If we return to our hypothetical scenario, we see that evolution might have been logically postulated, without any big jumps, when heritable continuous variation was shown to involve mutable Mendelian genes. And with the subsequent development of population genetics evolution would have become an inescapable inference.

The Unity of Biochemistry

Meanwhile, the study of comparative biochemistry provided impressive further support for macroevolutionary continuity. For although this field (like many novel developments in biology) was initiated simply to see what is there, it soon became apparent that the result fulfilled a firm evolutionary prediction: if all organisms have a common origin, then even after extensive divergence they might retain some common features.

Indeed, even by Darwin's time the microscope had shown a basic unity—all plants and animals are made of cells that share common structural features. But biochemical studies in the 1930s and 1940s revealed much more detailed evolutionary conservation. The most distant organisms, ranging from bacteria to man, use the same building blocks for their proteins and nucleic acids and, with variations, very similar enzymatic catalysts, central metabolic pathways, and energy-transducing mechanisms. Subsequently, organisms of all kinds were found to use the same fundamental molecular mechanism for regulating the activity of genes (and of enzymes), that is, allosteric changes in the shape of a protein. Clearly, a great deal of molecular unity underlies the morphological diversity of the living world. Unfortunately this evidence has penetrated very little into the introductory teaching of evolution.

More recently, molecular genetics has demonstrated a particularly dramatic unity—the genetic code. If organisms had arisen independently they could perfectly well have used different codes to connect the 64 trinucleotide codons to the 20 amino acids; but if they arose by common descent any alteration in the code would be lethal, because it would change too many proteins at once. Hence the finding of the same genetic code in microbes, plants, and animals (except for minor variations in intracellular organelles) spectacularly confirms a strong evolutionary prediction.

The Impact of Bacterial Genetics

The discovery of biochemical unity between bacteria and higher organisms encouraged the search for unity also in their genetic mechanisms. The result was the development of bacterial genetics, which then laid the groundwork for molecular genetics.

The emergence of bacterial genetics was curiously delayed. Changes in the properties of cultures during serial cultivation were one of the earliest challenges of the infant science of bacteriology; yet study of bacterial variation did not become linked to the new science of genetics, begun in 1900, for at least four decades. There were several reasons. The bacterial cell was too small to permit any chromosomal organization to be visualized, and genetic recombination could not be demonstrated (because it is a very rare event, rather than the regular event seen in the reproduction of higher organisms). Moreover, bacteria multiply so rapidly (up to three generations per hour) that the progeny of a highly favored mutant could replace a parental population during overnight growth; and since this shift seemed too rapid for a Darwinian process, a Lamarckian, instructed change in inheritance was inferred. Bacteria were therefore regarded as virtually bags of enzymes having some vague, plastic mechanism of inheritance.

The barriers were not broken down until the 1940s, when Beadle and Tatum's discovery of a 1:1 relation between genes and enzymes, in the mold Neurospora, was soon extended to bacteria. Even more significantly, Avery's study of pneumococcal transformation founded both bacterial genetics (by demonstrating gene transfer between bacteria) and molecular genetics (by showing that the genetic material is DNA). Bacteria and their viruses then became the favored models for the study of molecular genetics, because of their relative simplicity and rapid growth and because very rare mutants or recombinants could be efficiently selected from huge populations.

Although the presence of gene transfer and recombination in bacteria came as a surprise, in retrospect it is easy to recognize the evolutionary value of the emergence of these sophisticated processes in such early, simple organisms. Prokaryotes clearly had to accumulate an enormous amount of genetic variation during the 3 billion years before they gave rise to eukaryotes. If its only source were successive mutations within a lineage, without any form of recombination, evolution would surely not yet have progressed beyond early prokaryotes.

In addition to providing a background for molecular genetics, the development of bacterial genetics provided a simple demonstration of natural selection, in an experiment that can be easily performed in the elementary biology laboratory: the overnight emergence of resistant bacterial strains in cultures containing an inhibitor or a bacteriophage. (Moreover, in populations of humans occasionally treated with antibiotics resistant variants may become predominant after a few months or years).

Nine years after Avery's great discovery Watson and Crick launched the field of molecular genetics by elucidating the basic structure of DNA. The concepts and techniques that emerged have influenced virtually every branch of biology, and they have given rise to the rapidly growing daughter field of molecular evolution.[14,15] I shall briefly describe some particularly pertinent developments.

Size of Steps in Evolution*

As Watson and Crick noted, the structure of DNA not only explained the puzzle of gene replication: it also could explain mutations as inevitable occasional errors in base-pairing during replication. Further studies have revealed an extraordinary variety of additional molecular mechanisms of mutation, and the list is still growing. As a result, Darwin's thesis—that new organisms evolve by "numerous successive, slight modifications"—has acquired a precise, operational meaning. A "slight" modification now means a one-step hereditary change (a "mutation" in the broad sense).

In addition, the size of a mutation, originally described only in terms of phenotypic effect, can now be defined in terms of information content, and the two measures do not necessarily vary in parallel. In particular, some small changes in sequence can produce a large phenotypic effect—an obviously important process for evolution.

In the simplest mechanism, even a *single-base change* in one gene can have a broad effect, by affecting the products of a number of other genes. Base substitutions may have this effect when they occur in DNA sequences that regulate several genes, code for a topoisomerase, or code for an enzyme that can modify other proteins. Single-base changes can also cause a frameshift and thus result in a major change in the sequence of a protein.

---

*This rather condensed, technical section is not essential for the overall argument and can easily be skipped.

Another source of broad effects is *DNA rearrangements.* Microscopically visible chromosomal rearrangements that were discovered decades ago, but from McClintock's work on maize, and from more recent molecular studies in bacteria, it is clear that smaller rearrangements, invisible to the cytologist, are much more frequent. Moreover, they occur by highly evolved mechanisms, for they often involve mobile genetic elements (transposons), which have special terminal sequences and a gene coding for an enzyme that mobilizes these sequences. These transpositions not only can alter the quantitative expression of a gene, but they can create novel proteins by fusing parts of different genes: an obviously major evolutionary mechanism.

In eukaryotic cells *introns* (nontranslated intervening sequences between the translated parts of a gene) appear to serve to accelerate such recombinations of parts of different genes. Moreover, recombination also occurs in messenger RNA, and the resulting novel sequences can be copied into the DNA of the chromosome by reverse transcriptase. Since copying errors are much more frequent in RNA than in DNA (which has elaborate corrective mechanisms), it has been suggested that DNA guards genetic identity, whereas RNA promotes its modification.[16]

In higher organisms *embryonic development,* including differentiation and morphogenesis, can enormously amplify the effects of a genetic change. (For example, a rare single-gene defect in man—polydactyly—results in hands with six fingers). *Differentiation* is just beginning to be understood, for it depends on *selective gene regulation,* and several mechanisms have already been identified. These include binding of regulatory proteins, selective methylation of bases, shift of the DNA conformation to a left-handed twist (Z-DNA), rearrangements in DNA sequences, and selective gene amplification. *Morphogenesis* may prove to be much more obdurate, for it involves intricate patterns of product localization and feedback regulation in time and space, rather than the linear sequences of events familiar to biochemists today.

A large phenotypic effect can also be achieved in one step by the introduction of *exogenous DNA* into a cell—and often into a chromosome—whether by a virus, by a plasmid, or as a naked DNA fragment. Though such "infectious heredity" was discovered in bacteria (where it causes rare, partial genetic recombination in otherwise asexual organisms), it has also been observed in the cells of higher organisms. Indeed, this ability of a virus to become integrated into a chromosome, and then to transfer some of the host DNA into other organisms, has intriguing evolutionary implications. Like bacteria, which were first studied as pathogens and then found to have a much broader role in the recycling of organic matter, viruses may also have developed their prominent role in pathogenesis only as a sidetrack—in this case, from an evolutionary role in mediating gene transfer.[17,18]

Molecular genetics is thus producing profound insights into the mechanisms of genetic variation, and the end is not in sight. Such studies leave little justification for the recent speculation (reviving an earlier saltationism) that species formation requires some unknown kinds of macromutations, or for the further claim that the neo-Darwinian picture requires radical revision.[19,20] Moreover, these claims have been part of a broader argument, dramatized as "punctuated equilibrium."[19–21] This proposition—that evolution is not gradual but alternates between long periods of stasis and short bursts of rapid change—has been widely criticized for claiming a spurious novelty, since Darwin explicitly recognized that "gradual" meant proceeding through a sequence of small steps but not necessarily proceeding at a constant rate. In fact, Gould has recently reduced his claims so drastically[22] that the scientific controversy has evaporated. Unfortunately, reverberations in the press continue to contribute to the antievolution movement by creating the false impression that the neo-Darwinian view is in disarray.

Nevertheless, it is true that the detailed mechanism of speciation is still a challenge. The theory of evolution demands gradual change; yet we fail to find a continuous series of transition forms, both among fossils and (with few exceptions) among living species. The obvious explanation is that the intermediates in the formation of a new species quickly move forward in evolution into the better-adapted new species (like a transitory intermediate in a chemical reaction), and so their numbers are too low to permit detection. Moreover, although the neo-Darwinian model requires that the genetic steps between species form a continuous series, it does not specify the size of the individual steps, either in terms of information content or in terms of phenotypic effect—and we have seen that either can be larger than earlier seemed possible.

Developmental biology will surely play a major role in elucidating the steps in the transitions between species, since the genetic requirements for effective development must limit the directions in which an organism can evolve. Indeed, with further advances in developmental biology, and with our growing ability to sequence and to manipulate genes, it may become possible to reproduce the transmutation of species before long.

Verified Predictions, and the Origin of Life

Since evolutionary theory has been faulted by some philosophers for its presumed lack of testable predictions, it is interesting to note that studies in molecular genetics have verified several major predictions (though these were usually recognized as such only retrospectively). For example, in order for increasingly complex organisms to evolve the genome would have to be able to expand. One mechanism of expansion, already noted, is the insertion of a block of exogenous DNA. Another is nonreciprocal recombination (unequal crossover) between homologous chromosomes, which yields a chromosome with two copies of a particular sequence: one of these can then maintain the original function, while successive mutations in the other can create a new function. Confirming this evolutionary mechanism, many families of proteins with closely related sequences have been found within an organism.

Another prediction arises from the wide range of genome sizes, which is 10^4-fold from the smallest viruses to vertebrates. A mutation rate (per unit length of DNA) that is appropriate for evolution in the former group would create an excessive frequency of mutations (mostly deleterious) in the latter, and so the rate should show a roughly inverse correlation with genome size. This prediction has been confirmed. Three mechanisms that alter the rate are known: variations in (1) the accuracy of the enzymes that replicate DNA, (2) the activity of the enzymes that correct errors in replication, and (3) the presence of mobile sequences. In addition, the mutation rate can be varied at specific sites: for example, methylating a base increases its mutability without changing its specificity of pairing. This ability to select for altered mutability—in the whole genome or in a specific part—no doubt increases the flexibility of evolution.

Finally, another prediction is that a single DNA sequence can theoretically be read in multiple ways, and evolution might be expected to take advantage of the economy that would result. Three different mechanisms have been observed: (1) reading not one but both strands of a DNA sequence, (2) reading a strand between different starting and stopping sites, and (3) reading it in different phases (created by the fact that coding occurs in continuous triplets).

Molecular genetics has also provided an increasingly detailed answer to an old objection to Darwinian evolution: that it lacks a plausible beginning. Earlier speculations, by Oparin and Haldane, had postulated that organic molecules would accumulate in a prebiotic "soup" and that, in a dramatic event, their spontaneous aggregation would eventually form the first cell. However, with advances in the study of molecular replication it became clear that nucleic acids, as repositories of information, had to precede proteins in prebiotic evolution: hence natural selection could appear even before the first cell. As Eigen[23] has proposed, the evolution of life (or, more precisely, the evolution of genetic information) would start with spontaneously polymerized, short nucleic acid sequences that slightly favored the formation of similar sequences; and through natural selection, at a molecular level, sequences would gradually emerge with increased precision of replication and hence increased capacity to transmit complex information. Proteins would arise later.

The recent finding of enzymatic activity in certain nucleic acid sequences[24,25] dramatically supports this proposal.

Whatever the details of prebiotic evolution, it gave rise to cells very early: bacterial fossils have been found in rocks that were formed within the first billion years of the earth's 4.5-billion-year history. This speed suggests that life will inevitably evolve under the right chemical circumstances. The ultimate mystery is thus not the creation of life; it is the creation of a cosmos whose properties led to the evolution of life. And here we can again recognize different spheres for science and for religion: since science cannot provide an answer to the question of ultimate origins it has no conflict here with religious speculation.

Sequence Homology

Thus far we have seen that molecular genetics has supported the theory of evolution by propounding many mechanisms of variation, many testable predictions, and plausible mechanisms of prebiotic evolution. In an even greater contribution it has directly demonstrated, in two novel ways, the macroevolutionary linkage of present living organisms.[26]

The first kind of direct evidence is based on a strong prediction: if evolution has occurred through the accumulation of mutations, then as organisms diverge in phenotype they should similarly diverge in DNA (and in protein) sequences. Several techniques have been developed for estimating similarity of sequence: immunological cross-reaction between proteins (initiated with hemoglobin more than 75 years ago!); formation of hybrid double strands by mixing DNA from two species under appropriate conditions; and direct sequence determination of proteins or of specific DNA fragments. The results have confirmed many of the classical branching taxonomic trees, which were based primarily on morphological criteria. For example, two closely related, very recently evolved species, man and the chimpanzee, have an average sequence homology of approximately 99 percent[27] compared with 70 percent homology between the mouse and the rat.

On the other hand, the parallel between genotypic and phenotypic properties becomes a bit weaker under some circumstances. Thus, when an organism encounters an entirely new territory (or a major environmental change within its territory) it can undergo *adaptive radiation*, rapidly creating a variety of new species to fill different ecological niches in that territory. On the other side of this coin, species that fill similar niches in different territories may exhibit *convergent evolution*, that is, they may develop similar structures even though these have been derived via very different routes that have long been separated.

Studies on molecular sequences have revealed striking, previously unrecognized examples of these classical evolutionary mechanisms. Thus, the widely varied songbirds found in Australia were earlier classified as relatives of various Eurasian and North American genera, which they closely resemble; but DNA hybridization now shows that they are much more closely related to each other than to any birds on other continents (reviewed in [28]). What is surprising is the extraordinary extent to which the Australian adaptive radiation, starting from an immigrant bird 65 million years ago, has produced phenotypes convergent with the various birds found in Eurasia. This finding emphasizes that even though chance variations make evolution possible, the environment closely guides its course.

Incidentally, this exception helps prove the rule. A skeptic might claim that the observed correlations between phenotype and DNA sequence simply reflect the nature of the material with which the Creator worked, rather than a predictable consequence of evolution. But he would then have to explain why the Creator developed different correlations on different continents.

The study of DNA sequences has opened up a wide variety of theoretical problems with implications for evolutionary theory. We have already seen that a remarkably small set of changes (as in the chimpanzee and man), presumably located primarily in genes regulating development, can cause large differences in morphology and behavior. Another discovery has proposed a "molecular clock," based on the surprisingly constant (and characteristic) rate at which various proteins fix neutral mutations (i.e., base substitutions that generate a synonymous codon or that code for a functionally equivalent amino acid). Moreover, studies of DNA in eukaryotes have revealed many unexpected properties. These include the presence of enormous amounts of highly repetitive, untranslated DNA; "pseudogenes" that are almost identical with active genes but are not expressed; untranslated sequences (introns) interrupting the translated regions; reverse transcription from RNA to DNA; and localized recombination in somatic cells in the formation of antibodies (and very likely also in genes with other functions).

It is clear that selection acts not only on genes, individual organisms, and populations, but also on variable properties of the DNA itself. These include stability; the effect of the ratio of A-T to G-C base pairs on the tendency to "breathe" more readily in a given region (i.e., to expose the bases for external pairing); the presence of repeat sequences that permit internal pairing within a strand (thus forming a cloverleaf); the choice among synonymous codons; the presence of sequences that favor interaction with plasmids, viruses, and restriction enzymes; and the content of mobile sequences.

In particular, the presence of much apparently nonfunctional DNA presents a challenge. This category may include duplicated genes undergoing a shift from one function to another—or, more broadly, a reservoir of unused genes for evolution to "tinker" with.[29] Since the amount of this DNA is extraordinarily large, it has alternatively been suggested that much of it may be "selfish" or parasitic, accumulated simply because such opportunism is built into the mechanism of DNA replication.[30,31]

DNA Transfer across Species Barriers

In addition to DNA sequence homology, a second kind of direct molecular evidence for genetic continuity is the transfer of DNA between distant organisms—observed not only in the laboratory but also in nature. For example, the bacterium *Agrobacterium tumefaciens* initiates crown gall tumors in plants by integrating a tumor-inducing gene from a bacterial plasmid into a host cell chromosome.[32] In another example, a symbiotic luminescent bacterium has evidently taken up a gene from its host, the ponyfish, since the bacterium has the characteristic animal form of the enzyme superoxide dismutase (containing Cu or Zn) rather than the quite different form (containing Fe or Mn) found so far in all other bacteria.[33]

In such lateral transfer the gene might be expected to persist more often when the donor and the recipient are closely related, but because the differences between native and foreign genes would then be much smaller the transfer would be harder to recognize. However, DNA sequencing may offer a method of detection. Thus, in a pair of sea urchin species long separated in evolution the gene for one histone seems to have been shared quite recently, since its sequence showed less than 1 percent as much divergence as that observed in other proteins, including other histones.[34] It therefore seems reasonable to infer a continual slow flow of bits of DNA between species.

Recognition of this flow of DNA is not entirely an esoteric matter: it is quite pertinent to the recent controversy over the hazards of recombinant bacteria. Most of the debate failed to recognize the virtual certainty that the human species has been continually exposed to natural recombinants, such as intestinal bacteria that had taken up DNA released from lysed adjacent host cells. Accordingly, the classes of recombinant bacteria being produced in the laboratory are neither as novel nor as threatening to our survival as was initially assumed. In addition, the successful spread of an organism depends, as we have seen, on a coherent genome rather than on a particular powerful gene; hence the introduction of random DNA of distant origin would almost always impair survival. Wider awareness of these considerations—among biologists as well as laymen—might have spared much unnecessary anxiety.

Conclusions

In summary, although Darwin presented massive evidence for natural selection in *The Origin of Species,* he lacked evidence—and even a plausible mechanism—for an equally crucial part of his theory: the assumption of continual, limitless hereditary innovation. Accordingly, his theory could not fully meet the criteria customary in the experimentally based sciences (although for that reason the intellectual achievement was all the greater, broadening as it did our concept of the scientific method). With the discovery of genes and mutations and then the extension of Mendelian genetics to polygenic traits, genetics provided the hereditary mechanisms required for evolution, and the later development of population genetics made it possible to demonstrate microevolution, based on chance variation plus selection. Moreover, the extrapolation from microevolution to macroevolution was a logical necessity, supported by the graded homologies of present living groups of organisms, by an expanded, precisely dated paleontological record, and finally by the demonstration of extensive biochemical features shared by all organisms.

Today, microbiology and molecular biology have provided major new perspectives in evolutionary biology. This field was concerned first with identifying the sequences and the kinetics of evolutionary change, but we can now explore the detailed underlying mechanisms. The results have shown, first of all, that a variety of sophisticated molecular mechanisms of hereditary change (including genetic recombination) evolved very early, in the primitive bacteria. Moreover, many different mechanisms for generating mutations have been identified, and some have large phenotypic effects—a property that is obviously useful for effecting macroevolution. But an even greater contribution to evolutionary biology is the remarkably direct new evidence for macroevolution. This evidence is of two kinds: divergences in DNA sequences parallel divergences in phenotype, from the simplest to the most complex organisms; and small blocks of DNA occasionally move between species, reinforcing their genetic continuity.

To be sure, since evolution involves an inordinate number of variables, with complex interactions, we cannot predict its future course in detail. But this limitation does not undermine the scientific rigor of the theory, any more than the inability of meteorologists to predict the weather shakes our confidence in the underlying physical principles. Accordingly, the term "theory of evolution," with its overtone of tentativeness, is outdated: evolution is a fact, and evolutionary theory (quite a different concept) has a position in biology comparable to that of atomic theory in chemistry.

L'Envoi: Life and Information

In closing, we should note that molecular genetics has brought great unity to biology, through the development of concepts and techniques that now permeate many fields. In addition, it has provided a rigorous, marvelously simple answer to one of the deepest questions in biology: what fundamentally distinguishes living from nonliving matter? Earlier vitalists vainly sought novel forces, and more recently some physicists sought novel physical laws. However, we now see that the uniqueness of life does not derive from special physical forces or laws. It lies, instead, in the organization of its materials in a way that generates a unique property: *the storage of information within molecules.*

The concept of molecular information arose from studies on nucleic acids, which store such information in *sequences.* However, the concept applies as well to allosteric proteins, which store information in their *conformation.* They sense concentrations of a metabolite or a hormone, by changing their conformation when they bind it—and that information then regulates the activity of specific genes, enzymes, or cell-membrane components. This molecular mechanism, which developed early in bacteria, later evolved in animals into the nervous system, where the transduction of sensory stimuli, conduction along nerve fibers, and synaptic transmission all depend on allosteric proteins.

The interactions between allosteric proteins and genes also have epistemological implications, for they demonstrate concretely a continuity, both evolutionary and functional, between two kinds of knowledge: inherited knowledge, programmed in our genes and expressed through development (including that of the brain), and learned knowledge, acquired by experience and programmed largely in our brain. This insight tells us that Kant was right, and the naive empiricists were wrong. We start our lives with a great deal of inborn information, which is a priori for individuals but a posteriori as a product of evolution, and we then add a second class of learned information.

Evolution ties together the extraordinarily coherent, esthetically satisfying, and practically valuable body of knowledge that we have acquired about the living world, ranging from bacteria to man. And these insights do not destroy our sense of awe, as is often charged; rather, they shift the focus. For the miracle is not simply the wondrous complexity and beauty of the firmament, the living world, and man; rather, it is the ability of evolution to produce a creature that can learn to understand so much about these matters. Yet we are surrounded by a sea of unbelievers—including many highly educated people.

The problem of teaching evolution effectively is exacerbated by the long separation of two main streams in biology: the naturalists, asking how the diverse organisms arose and are distributed, and the experimental physiologists, asking how they function. Molecular genetics has brought these streams together in principle, but so far, the occupants eye each other warily. The evolutionists warn of excessive reductionism, and the molecular geneticists are turned off by the metaphysical verbiage at the fringes of the evolutionary literature. Perhaps it will take a new generation of graduate students, exposed to both disciplines, to consummate the intermarriage. Meanwhile we will surely improve our teaching of evolution, even at an elementary level, by building on the molecular evidence.

NOTES AND REFERENCES

1. T. Bethell, "Darwin's Mistake," *Harper's* (February 1976).
2. K. R. Popper, *The Logic of Scientific Discovery* (London: Hutchinson, 1959).
3. K. R. Popper, *Dialectica* 32 (1978):344.
4. K. R. Popper, Letter, *New Science* 87 (1980):611.
5. Despite Popper's brilliance in analyzing the scientific method with physics as a model, his understanding of biology exhibits serious limitations, including a failure to recognize the significant elements of refutable prediction in Darwin's picture (e.g., correlation of age of fossils with their stage of evolution). I shall not deal here with these issues, which have been dissected well by a philosopher especially interested in biology: M. Ruse, "Karl Popper's Philosophy of Biology," *Philosophy of Science* 44 (1977):638.
6. *New York Times* (August 29, 1982).
7. In a gratifying exception, a recent brochure by a committee of the National Academy of Sciences has given serious attention to the molecular evidence: *Science and Creationism* (Washington, D.C.: National Academy of Science Press, 1984).
8. E. Mayr, *The Growth of Biological Thought* (Cambridge, Mass.: Harvard University Press, 1982).
9. W. B. Provine, *The Origins of Theoretical Population Genetics* (Chicago: University of Chicago Press, 1971).
10. E. Mayr, *Animal Species and Evolution* (Cambridge, Mass.: Harvard University Press, 1963).
11. T. Dobzhansky, *Genetics of the Evolutionary Process* (New York: Columbia University Press, 1970).
12. J. S. Huxley, *Evolution: The Modern Synthesis* (London: Allen & Unwin, 1942).
13. E. Mayr and W. B. Provine, eds., *The Evolutionary Synthesis* (Cambridge, Mass.: Harvard University Press, 1980).
14. G. A. Dover and R. B. Flavell, *Genome Evolution* (New York: Academic Press, 1982).

15. D. S. Bendall, ed., *Evolution from Molecules to Man* (Cambridge: Cambridge University Press, 1983).
16. D. Reanney, "Genetic Noise in Evolution?" *Nature* 307 (1984):318.
17. N. G. Anderson, "Evolutionary Significance of Virus Infection," *Nature* 227 (1970):1346.
18. D. Reanney, "Extrachromosomal Elements as Possible Agents of Adaptation and Development," *Bacteriological Reviews* 40 (1976):552.
19. S. J. Gould and N. Eldredge, "Punctuated Equilibria: The Tempo and Mode of Evolution Reconsidered," *Paleobiology* 3 (1977):115-151.
20. S. M. Stanley, *The New Evolutionary Timetable* (New York: Basic Books, 1981).
21. S. J. Gould, "Irrelevance, Submission, and Partnership: The Changing Role of Paleontology in Darwin's Three Centennials, and a Modest Proposal for Macroevolution," in *Evolution from Molecules to Men*, ed. D. S. Bendall (Cambridge: Cambridge University Press, 1983).
22. S. J. Gould, "The Meaning of Punctuated Equilibrium and Its Role in Validating a Hierarchical Approach to Macroevolution," in *Perspectives in Evolution*, ed. R. Milkman (Sunderland, Mass.: Sinauer, 1982).
23. M. Eigen and P. Schuster, "The Hypercycle: A Principle of Natural Self-Organization," *Naturwissenschaften* 64 (1977):541-565; 65 (1978):7-41, 341-369.
24. B. Bass and T. Cech, *Nature* 308(1984):820-823.
25. C. Guerrier-Takada et al., *Cell* 35 (1984):849-857.
26. A. C. Wilson, S. S. Carlson, and T. J. White, "Biochemical Evolution," *Annual Review of Biochemistry* 46 (1977): 573-639.
27. M. C. King and A. C. Wilson, "Evolution at Two Levels in Humans and Chimpanzees," *Science* 188 (1975):107-116.
28. J. M. Diamond, "Taxonomy by Nucleotides," *Nature* 305 (1983):17-18.
29. F. Jacob, "Evolution and Tinkering," *Science* 196 (1977):1161-1166.
30. W. F. Doolittle and C. Sapienza, "Selfish Genes: The Phenotype Paradigm and Genome Evolution," *Nature* 284 (1980):601-603.
31. L. E. Orgel and F. H. C. Crick, "Selfish DNA: The Ultimate Parasite," *Nature* 284 (1980):604-607.
32. P. Zambryski, M. Holsters, K. Kruger, et al., "DNA Structure in Plant Cells Transformed by *Agrobacterium tumefaciens*," *Science* 209 (1980):1385-1391.
33. J. P. Martin, Jr., and I. Fridovich, "Evidence for a Natural Gene Transfer From the Ponyfish to its Bioluminescent Bacterial Symbiont *Photobacter leiognathi*," *Journal of Biological Chemistry* 256 (1981):6080-6089.
34. M. Busslinger, S. Rosioni, and M. L. Birnstiel, "An Unusual Evolutionary Behavior of a Sea Urchin Historic Gene Cluster," *EMBO Journal* 1 (1982): 27-33.

11

Speculating on the Brain

Reviews of C. Sagan, *The Dragons of Eden* (New York: Random House, 1977) and J. Jaynes, *The Origins of Consciousness in the Breakdown of the Bicameral Mind* (Boston: Houghton Mifflin, 1977).

The excellent academic credentials of the authors would lead the reader to assume that these two books were written with a higher degree of balance and scholarly responsibility than I found. It goes without question that fantasy, playfulness, and poetry are valuable in writing on science for a general audience—as, for example, in the charming essays of Lewis Thomas. It is another matter, however, for scientists to discard relevant facts selectively for the purpose of dramatization or advocacy.

For many years the impact of science on our culture has been a cause for concern as well as pride. The success of its approach to the external world has diverted attention from man's inner life; the resulting understanding of nature has destroyed the supernatural foundations of our earlier moral consensus, without providing a substitute system; and the technological applications of science are increasingly seen to have costs as well as benefits. But while these problems are disturbing, it is also disturbing to see a reaction to them that is creating widespread suspicion of the value of scientific research and even of rationality. For if we were to abandon science and technology we would only intensify the economic crises of our densely populated world. And if we were to reject the scientific search for objective truth about nature, instead of recognizing the complementarity between such knowledge and value judgments, we would impair our ability to make effective judgments based on reality.

Hastings Center Report (April 1978):34.

The Responsibilities of Speculation

As a counter to this trend Sagan's book, *Dragons of Eden*, deserves praise. It portrays science as an exciting and valuable enterprise, and its focus on a Darwinian approach to the origin and nature of the human brain is a welcome corrective to current mystical approaches. The book's popularity is not surprising. Not only is Sagan a celebrity, through his association with exobiology and his television appearances; he also writes with clarity and charm. In addition, a wealth of imagery, a range of associations, and a variety of literary and philosophical quotations increase the book's readability for nonscientists.

Unfortunately, despite these virtues, the book is seriously flawed. It lacks, both in what it contains and in what it omits, the high degree of rigor and responsibility that Sagan's credentials as a scientist lead us to expect. Loose, naive, or scientistic statements abound: for example, that memory may be contained in particular molecules (an ambiguous, and in its most obvious interpretation an obsolete, statement); that the brain of the chimpanzee is suitably prepared for the introduction of language, though "not to quite the same degree" as that of man (not quite?); and that further investigation of neocortical activity in the fetus will provide the ultimate key to the solution of the abortion debate (*pace* the naturalistic fallacy).

Experts on neurobiology have criticized other details. But Sagan confesses to being an amateur in this area, and he admirably communicates his primary message about the evolutionary and the neurobiological basis for our emotions and for intelligence. Such lapses are therefore forgivable. I am more disturbed by his practice of inserting "I wonder" every few pages, giving the innocent reader the illusion of being in on the ground floor of a series of exciting discoveries. This practice violates a canon that every graduate student in science should learn in the course of writing his first paper: namely, that speculations are cheap. A scientist is encouraged to publish new ideas that derive from his data, or from an unusually novel insight or an ingenious theoretical analysis. But he forfeits respect if he appears to offer ideas well known in the field as his own.

Skimming the surface of evolution and neurobiology, Sagan comes up with many obvious and sometimes silly speculations. For example: "I wonder whether the ritual aspects of many psychotic illnesses could be the result of hyperactivity of some center in the R-complex" (p. 62); "I sometimes wonder if deodorants, particularly 'feminine' deodorants, are an attempt to disguise sexual stimuli and keep our minds on something else" (p. 69). I, in turn, wonder why the author, after a quite informative chapter on the different functions of the right and the left halves of the neocortex, offers the following near-tautology: "Might schizophrenia be what happens when the dragons are no longer safely chained at night; when they break the left-hemisphere shackles and burst forth in daylight?" (p. 99).

These errors of commission are irritating, but certain omissions are more serious. At the start the author sets a laudable goal: "A better understanding of the nature and evolution of human intelligence just possibly might help us to deal intelligently with our unknown and perilous future." But since the future begins, at each instant, with the present, we must ask: what aspects of the subject are most relevant to our present social problems, most clouded by preconceptions derived from a prescientific age, suffer most from misunderstanding of the limits of science? Which are most essential for a responsible scholar to face seriously? My own list would include three main areas, which Sagan ignores.

First, the chapter on "Genes and Brains" gives the author a fine opportunity to clarify misconceptions about gene therapy. The aim of this field (still far off) is to learn how to replace the single defective genes that are responsible for such crippling hereditary diseases as phenylketonuria and sickle cell anemia. But though this is surely an admirable goal there is wide public fear that its achievement would permit future governments to use the same techniques to blueprint personalities.

In fact, however, the gap between single-gene replacement and blueprinting of personalities is enormous. Though we do not know in detail how the developing embryo translates genetic information into the wiring diagrams and the functional connections (synapses) of the brain, we can be sure that a very large number of genes must be involved in the process. In addition, we know that no genetic manipulations will be able to rewire the brain *after* it has been formed. This information seems to be more useful to the public than anything else one could convey about genetics and the brain. But Sagan entirely avoids the controversial subject of genetic engineering. Instead, in a chapter on "The Future Evolution of the Brain" he discusses such trivia as the future surgical implantation in our brains of small computer modules that will provide fluent knowledge of exotic languages.

Second, the author gives a good deal of attention to the implications of evolution for our social behavior, and he speculates extensively about the behavioral influences of the deeper layers of our brain (the reptilian complex and the limbic system), phylogenetically inherited from earlier vertebrates. This is the subject matter of sociobiology, a recently defined and controversial field that has been receiving a great deal of public as well as scientific attention. Yet the word "sociobiology" does not even appear in the book.

A third omission concerns the implications of evolution for human genetic diversity in behavioral traits. In recent years this diversity, and particularly the heritability of intelligence, has been the focus of intense controversy, in which questions of scientific fact have been excessively mingled with sociological concerns. This topic is surely an indispensable part of a serious discussion of the evolution of human intelligence. In omitting it Sagan misses an opportunity to clarify important aspects that have been eclipsed in the emotional debate.

What we have, then, is a portrait of science rather like a Norman Rockwell painting—upbeat, cheerful, enthusiastic, without a wart or wrinkle, without a whiff of controversy, and with serene confidence in the future. Yet the general topic of the book—evolution and intelligence—presents one of the most provocative intersections of science and society.

In Sagan's defense, one might grant that he chooses a more cosmological perspective. But the disappointing observations that he does offer on social problems suggest another conclusion: that however much his concern with hypothetical organisms in space has led to a deep interest in biology, it has not led to deep insights into the implications of biology for the nature of our species and for its problems on this planet.

Science or Metaphor?

In contrast to Sagan's surface analyses, Jaynes's book, *The Origins of Consciousness in the Breakdown of the Bicameral Mind,* appears at first sight to be a highly original, scholarly contribution. Instead it turns out to be literary fantasy with a dash of apparent evidence from neurobiology, achieving originality at the cost of common sense. Moreover, the book does not deal with the origins of consciousness; it is about the cultural evolution of attitudes toward consciousness.

Jaynes's thesis is based on a peculiar quality that he notes in the *Iliad* and the Bible: gods speak frequently to men, but there are no references to human introspection. From this record he concludes that consciousness did not develop until a later stage in civilization. He infers that it arose not through changes in cultural attitudes alone (that is, in the culturally generated, learned information stored in the brain) but through "neurological reorganization" (p. 374) in the relations between the two cerebral hemispheres, dependent on the development of writing. He even speculates that there has since been evolutionary selection against people who hear voices, leaving us now with a rather small number of such people. In fact, however, our knowledge of population genetics shows that such rapid selection, in a few dozen human generations, would be impossible.

Building on this hypothetical reorganization of the brain, and on the evidence of Roger Sperry and others on the functional differentiation of the right and the left hemispheres of the neocortex, he describes poetry and music, religion, and hypnosis as vestiges of the "bicameral mind" that all people possessed a few millenia ago. Moreover, ignoring any distinction between the presumed hallucinations of a hero of the golden age and the incapacitating hallucinations of a modern schizophrenic, he scolds psychiatrists for using drugs that decrease hallucination.

The overall argument is preposterous. The early poets may have invoked divine sources simply in metaphoric terms. Moreover, in an age that lacked any scientific basis for relating mind and matter, and that endowed every tree with a spirit, people could easily have also endowed their self-aware mental processes with a separate identity.

But neurobiology has now taught us that function is indissolubly linked with structure. Hence however little we understand consciousness in neurobiological terms, it is clearly an expression of the activity of a brain with a certain complexity of structural organization. Moreover, the human brain has not increased in size in the past hundred thousand years. So even if a Homeric hero ascribed his more profound thoughts to an external, anthropomorphic source, was he not also *conscious* of his more ordinary thoughts and decisions—to eat a meal, to pick up a weapon? And what about the peoples who have not achieved written tradition even up to modern times? Do they lack consciousness?

If this book were simply an exercise in untrammeled literary imagination one could admire the author's creativity and erudition. And some may admire his highly personal literary style. For example, "The yearning for certainty which grails the scientist, the aching beauty which harasses the artist, the sweet thorn of justice which fierces the rebel from the eases of life. . . ." (pp. 8-9). But he has chosen a title that implies a focus on a real scientific problem; he is a member of the Psychology Department at Princeton; and he invokes many scientific sources as well as literature and history.

The work, however, is in no sense a scientific contribution, popular or serious. Instead its propositions are in the tradition of classical theological tracts, expressed in terms that the reader may interpret either literally or metaphorically. Moreover, while the author goes through the motions of scientifically testing his hypotheses against reality, he ignores most of the reality encountered in everyday life.

Hemispheres of the Brain

Our present knowledge of the functional differences between the two halves of the cerebral cortex is provocative, but still primitive. Eventually this field will shed light on human behavior, just as recognition of the evolution of the limbic system helps us to understand emotional responses and social patterns that we share with lower mammals possessing a similar structure. But if we can say at present that the right hemisphere is to some degree more involved with imagination and with patterns and the left with language and logic, we must recognize that the correlation is loose. Moreover, there are no evolutionary antecedents to enrich our psychological insights, and there is no neuro-anatomical evidence on individual differences in the circuitry of the two hemispheres (though they no doubt exist). Hence substitution of neuroanatomical terms as synonyms for familiar behavioral concepts has no predictive value and gives only the illusion of adding depth to the discussion. Worse, the illuson tempts the author into excessive extrapolations, orbiting out of the range of the intellectual gravity that holds science together.

12

Ethical Aspects of Genetic Intervention

The Rev. Joseph Fletcher has been prominent in advocating a "situationist" approach to ethics—a pragmatic approach that has appeal for many scientists. My editorial comments on an article of his, in the same journal, that focuses on ethical aspects of genetic intervention. Other aspects of that subject are considered at length in Part Six of this book.

The dramatic recent advances in molecular and cellular genetics have led to the widespread belief that we will soon be able to cure hereditary diseases by genetic manipulation, and that the same techniques could then be used to control human behavior. Though these assumptions are unlikely to be true, for reasons discussed below, concern over the possible political abuse of such powers has clearly contributed to (and been fed by) the current wave of anti-science. Indeed, fear of genetic manipulation seems to have generated especially strong anxiety. Thus, a recent report of the highly responsible Friends Service Committee,[1] on moral aspects of medical progress, presented open-minded views on abortion and on the useless prolongation of physical life, but the chapter on genetics unfortunately opposed gene therapy on the grounds that it would set a dangerous precedent. Clearly, more public education on the technical realities of the problem, and widespread discussion of its moral and social aspects, are highly desirable.

It is therefore appropriate that physicians should be exposed to the

Extract from editorial in *New England Journal of Medicine* 285 (1971):799-801.

ideas of theologians. But it is clear that with such novel problems, and in a time of moral flux, we cannot expect professional moralists to agree on a prescription: the solutions, as with contraception and abortion, will be reached only through extensive debate. Thus, in a recent issue of the *New England Journal of Medicine* Paul Ramsey[2] presented a conservative view: his a priori, metaphysical conviction of the sanctity of the earliest conceptus, with its "surprising uniqueness and individuality," led to the insistence that we identify abortion with homicide. His parallel views on genetic intervention in man have been presented elsewhere.[3] In contrast, in this issue (page 776) Joseph Fletcher, from the same branch of organized Christianity, adopts a pragmatic view, emphasizing the relevance of risk-benefit calculations rather than a priori principles of supernatural or intuitive origin: moral judgment of any act should be based entirely on its consequences. It would follow that no act is good or bad in and of itself. This view parallels that of the student of evolution looking at the extremely pragmatic pressures of Darwinian natural selection: a gene is not good or bad in the abstract, but its value depends on how it functions within the genetic context and the environment of the individual.

For most people educated in science the pragmatic approach is probably more attractive than the metaphysical one, and the allegedly revealed basis of ethics seems generally to be losing ground. Indeed, the lack of a widely agreed-on alternative basis may be a major factor in much of our social unrest and violence today. Nevertheless, few are likely to be quite as radically pragmatic as Dr. Fletcher. It therefore seems worthwhile to call attention to his brief discussion of a third basis of morality, "rule-utilitarianism." In this approach categorical generalizations, derived empirically from observing the consequences of many acts, are trusted as having a firmer base than the attempt to predict the consequences of individual acts. As he points out, this approach underlies most of the current opposition to the use of genetic intervention in man, whether for therapeutic or for other purposes.

In response to those who would cut off advances in our knowledge, rather than face the consequent responsibilities, Dr. Fletcher argues that genetic intervention would not create slavery but would follow it, since such intricate procedures would surely require cooperation, forced or otherwise. We might expand this point by noting that the technical possibility of using genetic manipulation to promote tyranny is still remote, whereas other powerful tools are already at hand—psychologic, pharmacologic, and even neurosurgical. Moreover, we have already seen an ancient method of genetic control, selective breeding, barbarically applied to man in the form of genocide, with technology involving gas chambers or bombers, and not requiring sophisticated genetics. If we do not manage to confine these genies to their bottles it will surely be hopeless to try to protect ourselves against future evil by limiting progress in medical genetics.

Dr. Fletcher makes the further point that as our knowledge of genetics increases we will surely not accept indefinitely the "invisible hand" of blind natural chance in genetics, any more than we exempt from intervention such "natural" disasters as the cycles of laissez-faire economics or epidemics of smallpox. However, he does not consider the large difference between the physician's moral duty to prevent or cure specific hereditary diseases and the much broader mandate of the eugenicist, aiming at improving the gene pool—whether by decreasing the births of those with serious genetic limitations or by increasing the supply of those with special talents. Thus, we can readily agree that the gene for phenylketonuria is as undersirable as the germ for smallpox, but the value judgments involved in nonmedical eugenic decisions will not be so easy to agree on. One therefore wishes that Dr. Fletcher had discussed the conflicting interests and values that lie at the heart of ethical problems. One also wonders that he is so optimistic about our ability to control new technologies wisely: the violent current reaction of literary intellectuals and of idealistic youth against science and technology bears witness to the failures around us.

His defense of genetics against the neo Luddites could be strengthened by a third argument: that the scientific method cannot be unlearned. Moreover, curiosity and the drive for power and for control over our destiny provide powerful motives for both the use of this method and dissemination of the results. So whether human genetics advances faster or slower, and whether the knowledge arises here or in another country, we will surely continue to amass whatever knowledge can be cumulatively pried from Nature's storehouse. We might therefore do better not to try to run away from the linked threat and promise of discovery, but rather to encourage our society to try to learn how to increase the ratio of benefits to costs. And on complex and novel issues Dr. Fletcher's firmly pragmatic approach may prove inevitable. However much we try, society may well not be able to reach a firm position until the results of the first experiments are in.

NOTES AND REFERENCES

1. *Who Shall Live? Man's Control over Birth and Death*, report prepared by the American Friends Service Committee (New York: Hill and Wang, 1970).
2. P. Ramsey, "*The Ethics of a Cottage Industry in an Age of Community and Research Medicine*," *New England Journal Medical* 284 (1971):700-706.
3. P. Ramsey, *Fabricated Man: The Ethics of Genetic Control* (New Haven, Conn.: Yale University Press, 1970).

Part Three

Genetics, Racism, and Affirmative Action

13

Social Determinism and Behavioral Genetics

This editorial was stimulated by some remarkable statements by Richard Lewontin on a "NOVA" program aired by the Public Broadcasting Service. It was the beginning of my debate with the Science for the People group over their campaign against the study of human behavioral genetics—a movement that I later began to call neo-Lysenkoism.

The history of this publication may shed light on the atmosphere that has surrounded discussions in this research area, and it will explain why the piece refers coyly to "a distinguished geneticist," who was actually Professor Lewontin. Since he is Alexander Agassiz Professor at Harvard, and since the famous earlier Harvard biologist Louis Agassiz had been a leading opponent of Darwin, I originally titled the piece "Agassiz Returns—Stage Left." As a courtesy to a colleague, I sent Professor Lewontin a copy of the manuscript, whereupon he demanded that the journal give him equal editorial space for a reply. The Science editor properly refused to use the editorial page for such an exchange; but instead of insisting that Lewontin settle for his right to publish a critical letter, he declined to print my piece. The following editorial was finally accepted, however, when I agreed to eliminate any identification of Professor Lewontin.

The fusion of evolutionary theory with genetics has yielded several profound insights into the nature of man. We now know that most traits are determined by interaction between genes and the environment, rather than by either acting independently. Moreover, the traditional

---

Science 189 (1975):1049.

view of race, as a set of stereotypes with minor variations, has been invalidated by the knowledge that races differ statistically and not typologically in their genetic composition. Finally, the rapid evolution of our species implies wide genetic diversity, with respect to behavioral as well as to morphological and biochemical traits.

Unfortunately, the idea of genetic diversity has encountered a good deal of resistance. Some egalitarians fear that its recognition will discourage efforts to eliminate social causes of educational failure, misery, and crime. Accordingly, they equate any attention to genetic factors in human behavior with the primitive biological determinism of early eugenicists and race supremacists. But they are setting up a false dichotomy, and their exclusive attention to environmental factors leads them to an equally false social determinism.

Ironically, this opposition parallels that of theologians a century ago: both saw the foundations of public morality threatened by an implication of evolution. But neither religious nor political fervor can command the laws of nature. One might accordingly expect scientists, knowing this very well, to encourage the public to accept genetic diversity—both as an invaluable cultural resource and as an indispensable consideration in any approach to social equality. Yet in a recent "NOVA" program on the Public Broadcasting Service a distinguished population geneticist denied the legitimacy of human behavioral genetics, scorned the belief that musical talent is inherited and even minimized the contributions of genetics to agricultural productivity. Similarly, members of a group called Science for the People, criticizing a study of possible behavioral effects of chromosomal abnormalities, wrote of the "damaging mythology of the genetic origins of 'antisocial' behavior," as though one must choose between genetic and social causation rather than study their interaction.

To be sure, in behavioral genetics premature conclusions are all too tempting, and they can be socially dangerous. Moreover, even sound knowledge in this field, as in any other, can be used badly. Accordingly, some would set up lines of defense against acquisition of the knowledge, rather than against its misuse. This suggestion has wide appeal, for the public is already suspicious of genetics. It recognizes that earlier, pseudoscientific extrapolations from genetics to society were used to rationalize racism, with tragic consequences; and it has developed much anxiety over the allegedly imminent prospect of genetic manipulation in man. Hence one can easily visualize an American Lysenkoism, prescribing an environmentalist dogma and proscribing or discouraging research on behavioral genetics. But such a development would deprive us of knowledge that could help us in many ways: for example, to improve education (by building on the diversity of individual potentials and learning patterns), to decrease conflicts, to prevent and treat mental illnesses, and to eliminate guilt based on exaggerated conceptions of the scope of parental responsibility and influence.

In the continuing struggle to replace traditional myths by evolutionary knowledge the conflict over human diversity may prove even more intense and prolonged than the earlier conflict over special creation: the critics are no less righteous, the issues are even closer to politics, and guilt over massive social inequities hinders objective discussion. What the scientific community should do is not clear. At the least we might try to help the public to realize the value of scientific objectivity, separated from political convictions, in understanding human diversity. Long ago men began to understand chemical diversity when they gave up the search for a philosopher's stone, which they had hoped would transmute other elements into gold. Today in human biology we face a similar problem in learning to build on facts as well as on hopes.

14

Neo-Lysenkoism, IQ, and the Press

The preceding short article emphasizes my conviction that if we refuse to recognize the importance of genes for human behavioral diversity, and if we reject the use of science to help us understand and build on that diversity, our society will lose more than it will gain. The eminent geneticist Theodosius Dobzhansky presented the same view, more eloquently and at greater length, in Genetic Diversity and Human Equality *(New York: Basic Books, 1973).*

This view obviously did not prevail. The next decade saw an extraordinarily widespread denial of the pertinence of genetic differences, however obvious, for various educational and social problems, while virtually all those persons who disagreed were silent lest they be accused of racism. Arthur Jensen, a very able and responsible educational psychologist, collected massive evidence for the importance of genetic factors in intelligence, and for the high probability of statistical, but overlapping, differences between racial groups; and he predicted great harm to our educational system if we ignore that reality and attempt to legislate the facts of nature, instead of using those facts to help us maximize the opportunity for development in each individual, regardless of race. He was vilified as a racist, though his writings repeatedly emphasize that any differences between mean group values must not be used to justify discrimination against individuals.

Through this decade Stephen Jay Gould, though a member of Science for the People, had not been prominently involved in its attacks on human behavioral genetics. However, in 1981 he published a book entitled The Mismeasure of Man, *designed to give the coup de grace to all further controversy over the heritability of IQ. Previous critics had argued that*

The Public Interest (Fall 1983):41–59.

such measurements were methodologically unsound: but he sought to show that the very concept of general intelligence, which IQ tests aim to measure, had no meaning, and so its heritability would also have no meaning. Though the reviews of the book in leading scientific journals were highly critical, it was enthusiastically received by most of the general press, garnered a literary prize, and was clearly influential. I therefore decided to present my own critique and to discuss the reasons for the curious pattern seen in the reviews.

This essay also presents my views on human behavioral genetics, and on its politics, in more detail than elsewhere in this volume. The succeeding paper, a reply to Gould's rebuttal, will amplify this position.

Stephen Jay Gould, a professor of geology at Harvard, has become one of the best known American scientists. His many essays on natural history are entertaining and highly readable, and his attack on the "establishment" version of Darwinian evolution has received so much attention that his picture appeared on the cover of *Newsweek*. He personalizes his expository writing in a breezy, self-deprecating manner, and he comes across as warm-hearted, socially concerned, and commendably on the side of the underdog. Hence he is able to present scientific material effectively to a popular audience—a valuable contribution, and a public service, as long as his scientific message is sound.

It is therefore not surprising that Gould's history of the efforts to measure human intelligence, *The Mismeasure of Man*, received many glowing reviews in the popular and literary press, and even a National Book Critics Circle award.[1] Yet the reviews that have appeared in scientific journals, focusing on content rather than on style or on political appeal, have been highly critical of both the book's version of history and its scientific arguments. The paradox is striking. If a scholar wrote a tendentious history of medicine that began with phlebotomy and purges, moved on to the Tuskegee experiment on syphilitic Negroes, and ended with the thalidomide disaster, he would convince few people that medicine is all bad, and he would ruin his reputation. So we must ask: Why did Gould write a book that fits this model all too closely? Why were most reviewers so uncritical? And how can nonscientific journals improve their reviews of books on scientific aspects of controversial political issues?

Reviews in the Popular Press

Typical of the literary reviews of Gould's book is the one that appeared in the *New York Times Book Review*. June Goodfield, a historian and popular writer on science, is effusive: In his "most significant book yet,

Mr. Gould grasps the supporting pillars of the temple in a lethal grip of historical scholarship and analysis—and brings the whole edifice of biological determinism crashing down." *The Mismeasure of Man*, she writes, also shows that, while science can never be wholly objective, "this gloriously human enterprise does provide us both with a method for challenging the status quo and for revealing true knowledge about the world." Moreover, Gould "affirms that most things are humanly possible, and that attempts to confine human beings to limited categories are both downright wicked and bound to be self-defeating."

In the *New Yorker* the book was reviewed by Jeremy Bernstein, a philosophically inclined physicist. His analyses of scientific books have in general been excellent, and we might have expected him to be critical of Gould's methodology. But in fact, because Bernstein saw the book as a powerful salvo against racism, he misread it, imputing to Gould his own, different views on intelligence. Bernstein's answer to racism is to emphasize "how numerous the genetically expressed variations are *within* any social group," whereas Gould in fact insists that in the area of behavior, genetic differences should be ignored. Missing this fundamental disagreement, Bernstein uncritically accepts Gould's indictment of intelligence tests: "because of the false reification of intelligence hundreds of thousands—perhaps millions—of people's lives have been circumscribed or even ruined."

The most perplexing review is Richard Lewontin's in the *New York Review of Books*. Lewontin represents a biased choice on the part of that journal, since he and Gould had taught a course together at Harvard on the dangers of applying biology to society, and he has called for the development of a true "socialist science" to challenge the "bourgeois science" of most Western culture. Yet he turns out to be an interesting choice, for his article is, as usual, brilliant, erudite, and idiosyncratic.

Lewontin agrees that political views, whether good or bad, will inevitably influence the conclusions of scientists, but he chides Gould for ignoring Marxist principles and overemphasizing racism: "*The Mismeasure of Man* remains a curiously unpolitical and unphilosophical book." The emphasis "on racism and ethnocentrism in the study of abilities is an American bias." Further, "In America, race, ethnicity, and class are so confounded, and the reality of social class so firmly denied, that it is easy to lose sight of the general setting of class conflict out of which biological determinism arose." He concludes with a profoundly pessimistic bit of metaphysics: "The reification of intelligence . . . is an error that is deeply built into the atomistic system of Cartesian explanation that characterizes all of our national science. It is not easy, given the analytic mode of science, to replace the clockwork mind with something less silly." But "the wholesale rejection of analysis in favor of an obscurantist holism has been worse. Imprisoned by our Cartesianism, we do not know how to think about thinking." It is unfortunate that this truly gifted scientist trapped himself in evolutionary genetics, a field so at odds with his social convictions.

The popular press has thought the issues to be more clear-cut. *Newsweek* refers to "this splendid new case study of biased science and its social abuse." *The Saturday Review* speaks of "a rare book—at once of great importance and wonderful to read." *The Atlantic Monthly* says, "The tale would be funny if one could overlook the misery that such tests have inflicted on generations of defenseless school children." *The Key Reporter* (of Phi Beta Kappa) calls the book "a strident, polemical, effective critique."

The Scientific Reviews

While the nonscientific reviews of *The Mismeasure of Man* were almost uniformly laudatory, the reviews in the scientific journals were almost all highly critical. In *Science*, a widely read American publication that covers all the sciences, the book was reviewed by Franz Samelson, a psychologist at Kansas State University. He concludes that as a history of science the book has a number of problems. For example, he notes, Gould claims that Army intelligence tests led to the Immigration Restriction Act of 1925; in fact, *no* psychologist testified before Congress, and the three reports of the House Committee on Immigration do not mention intelligence tests at all. On another point, Gould's discussion of the "fallacy of reification"—the grouping of different abilities, such as verbal reasoning and spatial reasoning, into one measure of intelligence—"remains blurred, since Gould's emphasis seems to shift about. Exactly what does he object to? [Gould] never tells us directly what his own proper, unreified conception of intelligence is." Finally, Gould fails to acknowledge that ability testing is "a sizable industry in the real world and a smaller one in academia. And all Gould's incisive thrusts at finagling and fallacies seem to be almost irrelevant. . . . Whatever intellectual victories over the [mostly dead] testers Gould's eminently readable book achieves . . . the real action seems to be elsewhere."

In *Nature*, a distinguished British journal of general science, Steve Blinkhorn, writing from the Neuropsychology Laboratory at Stanford University, is blunt: "With a glittering prose style and as honestly held a set of prejudices as you could hope to meet in a day's crusading, S. J. Gould presents his attempt at identifying the fatal flaw in the theory and measurement of intelligence. Of course everyone knows there must be a fatal flaw, but so far reports of its discovery have been consistently premature." More specifically, "the substantive discussion of the theory of intelligence stops at the stage it was in more than a quarter of a century ago." Gould "has nothing to say which is both accurate and at issue when it comes to substantive or methodological points." Finally, many of his assertions "have the routine flavor of Radio Moscow news broadcasts when there really is no crisis to shout about. You have to admire the skill in presentation, but what a waste of talent."

Science '82, a journal designed for the general public, chose as its reviewer Candace Pert, a biochemist at the National Institute of Mental Health, who has been researching the application of molecular biology and cell biology to the study of the brain. "Gould's history of pseudo-scientific racism in measuring human intelligence," she writes, "does not, despite his claims, negate the sociobiological notion that differences in human genetic composition can produce differences in brain proteins, resulting in differences in behavior and personality." In her view, "if modern neuroscience reveals biochemical differences that account for human variability, we must deal with this important knowledge; . . . ignoring differences because they *could* become abuses will not make them go away."

The most extensive scientific analysis of Gould's book appeared in *Contemporary Education Review.* Arthur R. Jensen, of the Institute for Human Learning at the University of California, Berkeley, analyzes Gould's technical arguments in great detail and reaches sharply critical conclusions. He also discusses recent research demonstrating a high correlation of IQ with speed of information processing, as measured by simple reaction-time techniques. These findings encourage a hope that a merger with neurobiology may soon make studies of intelligence much more penetrating and less controversial.

The review that appeared in *Scientific American* is an exception to the harsh criticism in the scientific press. Ordinarily *Scientific American* presents solid science in a very interesting way to a broad audience, and it has been restrained and nonpartisan in treating most controversial issues of science. However, there is one exception: The publisher, Gerald Piel, and the book editor, Philip Morrison, have long seen the study of the genetics of intelligence as a threat to racial justice. According to Morrison, as "a persuasive chronicle of prejudice in science, founded on scrupulous examination of the record, enlivened by the talent of a gifted writer, this volume takes on some of the sinister appeal of a tale of heinous crime."

Gould's Selective History

It is important for the general public to understand why scientists close to the field have reacted so negatively to *The Mismeasure of Man.* The strength of science in analyzing reality comes from its strict separation of facts from values, of observations from expectations. Measurements of intelligence, and of its hereditary and environmental origins, are part of natural science—even though one must go beyond science, bringing in judgments of value, in order to probe the social implications of the results. Hence any purported scientific exposition of these topics must be as dispassionate and objective as possible about the facts, whatever the social views the author favors. These are precious standards, whose corruption we must resist. Unfortunately, throughout Gould's book they are not met.

The early chapters describe in detail some extremely naive nineteenth-century attempts to measure intelligence in terms of brain size or body shape. These are fossils from the history of mental testing, and their excavation would ordinarily bore most readers. Gould, however, uses them skillfully, both to give the impression of a thorough scholarly analysis and to arouse indignation at such evil uses of science. Unfortunately, the advocacy and the emotional appeal betray the scholarship. In the early stages of any science, naive ideas, often reflecting the prejudice of the time, are inevitable. Gould infers that this legacy will persist; but history demonstrates that the advance of science depends on continually discarding false hypotheses and preconceptions. Gould further arouses the reader's indignation by describing the ill-informed and prejudiced views of Paul Broca and Louis Agassiz on racial differences. But at a time when slavery was legal, and long before the science of genetics revolutionized our understanding of the nature of race, it is hardly surprising that these views were held by leading scientists—and even, as Gould notes, by such enlightened social critics as Benjamin Franklin and Thomas Jefferson. To remind us of these roots in the history of racism is instructive—but to imply a similar prejudice in today's investigators of intelligence is unfair.

After emphasizing that Alfred Binet developed the first intelligence test, in France in 1905, only in order to improve the education of backward children, Gould goes on to describe misuses of the subsequent tests. His most horrifying example is a primitive study conducted in 1912, in which H. H. Goddard administered intelligence tests to a number of Ellis Island immigrants. He set his standards at an absurdly high level, classifying in the end an extraordinarily large percentage of subjects as "feeble-minded"—a term that then included "morons" who could nonetheless manage to make a living, though it is now applied only to those with a more severe deficiency. Probably nothing has so aroused antipathy to intelligence testing as his widely cited findings that, for example, 83 percent of the Jews and 79 percent of the Italians he tested were "feeble-minded."

Gould summarizes his interpretation of Goddard's findings as follows: "Could anyone be made to believe that four-fifths of any nation were morons?" But let us look at what Goddard actually wrote. The first sentence of his paper states that "this is not a study of immigrants in general but of six small highly selected groups," leaving out those at either end of the scale who were "obviously" either normal or feeble-minded.[2] At that time immigration officers were using subjective impressions to reject those people who appeared to be too retarded to learn to make a living, and Goddard hoped that tests could provide a more reliable basis for such decisions. Surprised at the results, he added a discussion that Gould conveniently ignores:

> Are these . . . cases of hereditary defects or cases of apparent mental defects by deprivation? . . . We know of no data on this point, but indirectly we may argue that it is far more probable that their condition is due to environment than that it is due to heredity. To mention only two considerations: First, we know their environment has been poor. It seems able to account for the result. Second, this kind of immigration has been going on for 20 years. If the condition were due to hereditary feeble-mindedness we should properly expect a noticeable increase in the proportion of the feeble-minded of foreign ancestry. This is not the case.

Goddard ended up *favoring* the immigration of people who appeared to possess limited present intelligence: Not only would they perform useful work, but "we may be confident that their children will be of average intelligence and if rightly brought up will be good citizens." Goddard was hardly a great scientist, but he deserves a fair hearing. The statements cited here hardly warrant Gould's conclusion that to Goddard "the cure [for feeble-mindedness] seemed simple enough: don't allow native morons to breed and keep foreign ones out."

After some years, as Gould notes, most of the early enthusiasts changed their views. Goddard, Terman, and Brigham each admitted that he had overestimated the ability of tests to detect innate differences and had underestimated the influence of cultural background. One might take this example of growth in understanding as a sign of the whole field's increasing maturity and objectivity. Gould, however, sees these confessions only as support for his accusation of bias.

What Is "Biological Determinism"?

Gould's own degree of bias is unusual in a work by a scientist. What is the source of this passion? Not mental testing itself, he makes it clear. Rather, his arguments against this testing are merely weapons for attacking the real enemy: what he calls "biological determinism."

As Gould correctly points out, early investigators who tried to measure intelligence were indeed *determinists:* They had the illusion that they were directly measuring a capacity determined by the genes. But while he continues to tar investigators of behavioral genetics with this brush, in fact they are now all *interactionists*. For while genetics necessarily began with the simplest relationships, in which a single gene determines a trait (such as the color of Mendel's peas, or a human blood type), the science eventually moved on to the quantitatively varying (metric) physical or behavioral traits, which socially are much more interesting. These were found to depend on multiple genes, and also on their cumulative interactions with the environment. This concept is now precisely formulated as the concept of *heritability:* a measure of what fraction of the total variance in a trait, in a particular population, is due to genetic differences between individuals—the other fraction coming from environmental influences.

Since Gould would prefer to combat the straw man of naive, "pure" determinism, he fails to note how the science of genetics has extensively replaced this concept with interactionism. But since he is too familiar with biology to deny this conceptual shift, he appropriates it for his own ideological argument: "The difference between strict hereditarians and their opponents is not, as some caricatures suggest, the belief that a child's performance is all inborn or all a function of environment and learning. I doubt that the most committed antihereditarians have ever denied the existence of innate variation among children." Curiously, "hereditarians" (Gould's misnomer for interactionists) are not credited with a similar appreciation of both factors. Instead, they are neatly skewered by being called "strict."

What, then, is the quarrel about? According to Gould, "the differences [between the camps] are more a matter of social policy and educational practice. Hereditarians view their measures of intelligence as measures of permanent inborn limits. Children, so labeled, should be sorted, trained according to their inheritance and channeled into professions appropriate for their biology." But good investigators, such as Binet, did not want mental testing to become a theory of limits. For them, Gould argues, "mental testing becomes a theory for *enhancing potential* through proper education [emphasis added].[3]

This is a deliberate effort to blur the issue. With one hand Gould concedes innate differences, and with the other he takes them away. If the two camps really differ mostly about social policy and not about the importance of hereditary factors, why does he struggle so to deny the latter? Similarly, whether the hereditary component is large or small, is it not a fact that individuals differ widely in their phenotypic, developed ability to absorb various kinds of education and to perform various kinds of jobs? Yet the book has not one word about the possible value of mental tests for educational and vocational placement or for comparing educational programs. (However, consistent with Gould's admiration for Binet's circumscribed aim, he does note the value of mental tests in guiding the therapy of his own child.) Finally, in describing the incredibly crude use of the Army's "Alpha" tests in 1917, Gould ignores the current use of sophisticated tests to help the armed forces select candidates for expensive training programs.

It is sad that Gould, preoccupied with the destructive social consequences of earlier biological misconceptions, is convinced that any modern studies on human behavioral genetics must have similar consequences. For to the contrary, modern evolutionary biology has had an opposite effect—by providing a powerful argument against racism. In the past, a widely accepted justification for race discrimination stemmed from a Platonic doctrine that prevailed for over two millenia: the belief that we can best understand groups of entities (including species and races) in typological (essentialist) terms, i.e., characterizing all the individuals in a group in terms of a hypothetical ideal type or essence, and dismissing differences from the ideal as trivial. Today, however, population genetics has shown that all species are genetically diverse, and that the differences are not trivial but rather are the source of evolution. With this shift from an essentialist to a populationist view, the genetic differences between races (except for some superficial physical traits) are now seen to be statistical rather than essentially uniform. And since the statistical distributions overlap extensively from one group to another, one cannot infer an individual's potential from his race.

If the pregenetic, typological misconceptions still prevailed, the modern revolt against race discrimination would surely have encountered much greater resistance, and it might even have been impossible. Unfortunately, biology has received little credit for this major social contribution, and none at all from Stephen Jay Gould.

The Concept of General Intelligence

The historical chapters, constituting most of *The Mismeasure of Man*, serve to convince the reader that the measurement of intelligence is immoral. But after this build-up, Gould, shifting from historian to scientist, offers an even sharper objection: The measurement is also unscientific.

The tests were originally developed for use by teachers, who often have trouble deciding whether a pupil's poor performance is primarily due to limitations in motivation or to limitations in ability. Their initial purpose, as we have noted, was to provide a more objective and reliable supplement to the teacher's subjective impression, in order to help pupils who are doing badly. But this early use of testing inevitably led to the development of additional possibilities. For example, by ranking the whole class, the tests also detected students who could move faster than the average. In addition, more specialized tests have evolved, especially for advanced students and for purposes of job placement. But as practical tools in public education, the most widely used tests are still composite ones, designed, like Binet's test, to cover a range of abilities pertinent to the whole curriculum.

Psychologists generally agree that the greatest success of their field has been in intelligence testing—both practical, in estimating individual abilities, and theoretical, in exploring the cognitive functions of the human brain. For it might have turned out that the determinants of different cognitive abilities were uncorrelated: that is, that the levels of abilities might be distributed independently. But in fact, tests for different kinds of intelligence—the ability to assimilate, retain, process, and express different kinds of complex information—show a remarkably high correlation in their results. The rank-ordering of most individuals is similar—but not identical—on a verbal test, an arithmetic test, or a nonverbal test involving spatial patterns. These results confirm an impression that we all tacitly build on in our daily lives: Some people are generally brighter than others, but people also differ in their special aptitudes. Both sets of differences are partly inborn and partly due to factors affecting the development of the inborn potentials.

The common factor shared in different cognitive abilities, as determined by statistical analysis of their correlations, was named g by Charles Spearman. In the ordinary IQ tests it contributes well over half the variance within a population, the rest representing uncorrelated differences in special abilities. Some day the basis for both kinds of variation will no doubt be better understood in cellular and biochemical terms. Indeed, it is encouraging that studies of the brain are rapidly progressing from its simpler integrative functions, such as the processing of visual stimuli, to more complex cognitive activities. Meanwhile, though, it is fruitful for psychologists to examine intelligence at the level of performance, and to compare ways of improving that performance, just as geneticists could usefully deal with genes as formal units long before discovering their molecular structure and mode of action.

Examined at this level, such tests have unquestionably helped innumerable teachers to identify pupils whose brightness was concealed by shyness, cultural barriers, or rebelliousness. On the other hand, there is also no doubt that the tests have often been interpreted or applied badly. If teachers focus excessively on general intelligence, measured on a one-dimensional scale, they may fail to encourage the development of each individual's particular strengths. Moreover, the assumption that g is entirely innate may persist in some quarters even though the concept of heritability (fractionation into genetic and environmental components) has now completely replaced that early view among scientists. But perhaps the greatest danger is that the test results may tend to be regarded as some kind of index of social worth, instead of recognizing that they measure only a limited set of behavioral traits. For while these are key traits for certain educational and vocational purposes, the tests ignore many other traits that also have great social value: for example, physical attractiveness, motor skills, creativity, artistic talent, social sensitivity, and features of character and temperament. The concept of any single scale of social worth has no meaning. Gould, however, keeps the reader's indignation alive by regularly defining the objective of the tests as the measurement of "worth"—sometimes qualified as "intellectual worth," but often unqualified, or even denoted as "innate worth."

Gould is clearly not interested in evaluating the past uses of intelligence tests fairly, or in improving their use. To him the tests must be extirpated because—and here we get back to the real villain—in using them to compare *individuals* one inevitably runs into consistent differences in the mean values for various racial and socioeconomic *groups*. "This book . . . is about the abstraction of intelligence as a single entity . . . invariably to find that oppressed and disadvantaged groups—races, classes, or sexes—are innately inferior and deserve their status."[4]

This statement, for all its hyperbole, captures what the book is about: Concerned with group differences, Gould has decided not to add to the polemics on their causes, but to attack the problem at another level. For if he can demonstrate that the very concept of measurable intelligence is meaningless, then it follows that all those disturbing data on group differences are meaningless as well. His weapon is his "discovery," first announced in the *New York Review of Books*, of two alleged "deep fallacies" underlying the concept of general intelligence: reification and the factoring of intelligence.

The "Deep Fallacies" of Reification and Factoring

Gould's argument on reification purports to get at the philosophical foundation of the field. He claims that general intelligence, defined as the factor common to different cognitive abilities, is merely a mathematical abstraction; hence if we consider it a measurable attribute we are reifying it, falsely converting an abstraction into an "entity" or a "thing"—variously referred to as "a hard, quantifiable thing," "a quantifiable fundamental particle," "a thing in the most direct, material sense." Here he has dug himself a deep hole. If this implication of localization is a fallacy for general intelligence, why is it not also a fallacy for specialized forms of intelligence, which Gould professes to accept? Going even further, he seems to abandon materialism altogether: "Once intelligence becomes an entity, standard procedures of science virtually dictate that a location and physical substrate be sought for it. Since the brain is the seat of mentality, intelligence must reside there." But we must ask what reasonable scientific alternative there is. A Cartesian dualism, in which mental processes exist apart from a material base?

Indeed, this whole argument is fantastic. The scientist does not measure "material things": He measures properties (such as length or mass), sometimes of a single "thing" (however defined), and sometimes of an organized collection of things, such as a machine, a biological organ, or an organism. In a particularly complex collection, the brain, some properties (i.e., specific functions) have been traced to narrowly localized regions (such as the sensory or motor nuclei connected to particular parts of the body). Others, however, depend on connections between widely separated regions. Accordingly, the *reality* of generalized intelligence—or equally, of any specialized cognitive ability—does not require a "quantifiable fundamental particle." Like information transfer in a telephone network or in a computer, cognition would be much the same whether the cells involved are grouped together in one region of the brain or are connected by fibers running between dispersed locations.

It is astonishing that a scientist with Gould's credentials, and with ready access to colleagues in the relevant fields, would present such a phony "discovery" as the fallacy of reification, and on the basis of truly antiquated views of neurobiology. He writes that the existence of general intelligence could have been proved correct "if biochemists had ever found Spearman's cerebral energy." This phrase refers to a particularly thin speculation, in the 1920s, about the physical basis for differences in IQ. But neurobiologists today simply do not deal in such vague concepts. Instead, they measure variation in the richness of cells, and connections, and neurotransmitter molecules in different areas of the brain.

The molecular studies linking these features of the brain to genes have hardly begun. But it is clear that this molecular biology must build on the principle that genes code for specific molecular components in brain cells, as in all other cells, and that these genes, like other genes, will vary from one individual to another. Moreover, these gene products in the brain will give rise to variation not only in its wiring diagram but also in the switches (synapses) that transmit impulses between its nerve cells. We are unlikely to be able to correlate intelligence with the incredibly complex and subtle circuitry of the brain for a long time to come; but it is not hard to imagine correlation with molecular differences in a class of synapses in different brains, affecting the speed of processing information just like differences in the transistors of difference computers.

Let us look at Gould's second "deep fallacy," factoring, is statistical.

Here he reconstructs an old controversy, which the field has long outgrown. In this dispute, Spearman calculated g (the measure of general intelligence) by running tests for different abilities and analyzing their correlations so as to extract their common component. Thurstone, whom Gould admires as "the exterminating angel of Spearman's g," preferred to focus on the *specialized* differences in intelligence. He therefore analyzed the results in a way that did not extract the overall correlation, but dispersed it among the differentiated primary factors. But the correlation did not disappear: Another calculation could extract it from the primary factors as a "second-order" g. Gould, however, sets out to "prove" mathematically that the primary correlation is a statistical artifact and that the second-order one is negligible.

To analyze Gould's unconvincing argument would be irrelevant. For in the end, after claiming to have disproved the correlations, he casually accepts them as self-evident: "The fact of pervasive positive correlation between mental tests must be one of the most unsurprising major discoveries in the history of science." This is itself a very curious judgment. In fact, the correlation is not inevitable or self-evident, for the brain might have been so constructed that a strong endowment of cells for verbal skills would leave less room for cells concerned with numerical abilities, etc. Different cognitive abilities might then exhibit no correlation, or even a negative correlation, and psychologists would then have found no general intelligence to measure.

Gould's arguments about g are irrelevant for another reason as well: Though he believes they support his aim of slaying the dragon of the heritability of intelligence, the assumed link to that problem does not exist. "The chimerical nature of g is the rotten core of Jensen's edifice, and of the entire hereditarian school. . . . Spearman's g, and its attendant claim that intelligence is a single, measurable entity, provided the only theoretical justification that hereditarian theories of IQ have ever had." This assertion is utterly false. *Whether an IQ test measures mostly general intelligence or mostly a collection of independent abilities, the heritability of whatever it measures will be precisely the same.* IQ's factor structure simply does not enter the equations for calculating its heritability.

It is unfortunate that Gould contrasts general and special intelligence with such overkill, for the differences deserve serious consideration, and the advance of behavioral genetics, focusing on units of inheritance, will force psychologists to aim for a more refined dissection of cognitive functions. But the prospect of such advances does not require us to deny that a wider, overall measurement has had historical value, and may still have practical value for educational purposes.

Objectivity in Science

In addition to moral and technical objections to mental testing, Gould offers an epistemological argument that has much broader implications: "I criticize the myth that science itself is an objective enterprise. . . . By what right, other than our own biases, can we identify Broca's prejudice and hold that science now operates independently of culture and class?" On the other hand, he adds that "As a practicing scientist, I share the credo of my colleagues: I believe that a factual reality exists and that science, though often in an obtuse and erratic manner, can learn about it." This is all very well—but throughout the rest of the book he proceeds as though objectivity is a myth and no factual reality can be discovered.

In fact, the key to the success of the scientific enterprise is its passionate dedication to objectivity: Its advance depends on accepting the conclusions dictated by verifiable observations and by logic, even when they conflict with common sense or with treasured preconceptions. To be sure, some years ago Marxist philosophers, generalizing from the influence of social and economic arrangements on many aspects of our behavior, initiated an attack on the objectivity of science. Moreover, this view has become rather widely accepted in the social sciences. But the study of the genetics of intelligence is a part of natural science, rather than of social science, even though its findings have relevance for social questions. If the science is well done it will tell us objectively what exists, without value judgments; only in the social applications of that knowledge will these judgments arise. For example, insights into the range and distribution of abilities do not tell us how much of our educational resources to devote to the gifted and how much to the intellectually handicapped; this knowledge simply improves our recognition of the reality with which we must cope.

The main source of confusion here is that the word "science" is used with three different meanings, in different contexts: science as a set of activities, as a methodology, and as a body of knowledge. The *activities* of a scientist certainly depend heavily on nonobjective factors. These include the resources and the incentives that a society provides for pursuing particular projects, and also the personal choice of problems, hypotheses, and experimental design. The *methodology* of science is much more objective, but it is also influenced by fashions in the scientific community. The *body of scientific knowledge,* however, is a very different matter. Its observations and conclusions, after having been sufficiently verified and built upon, correspond to reality more objectively and reliably than any other form of knowledge achieved by man.

To be sure, attachment to a cherished hypothesis may lead a scientist into error. Moreover, at the cutting edge of a science contradictory results and interpretations are common. But the mistakes are eventually discarded, through a finely honed system of communal criticisms and verification. Thus Broca's name has been immortalized by its assignment to a structure in the brain that he recognized, whereas his premature efforts to correlate gross structural variations with intelligence have left no residue in the body of scientific knowledge.

Accordingly, however much the findings in some areas of science may be relevant to our social judgments, they are obtained by a method designed to separate objective analysis of nature from subjective value judgments. Long experience has shown that when these findings are well verified, they have an exceedingly high probability of being universal, cumulative, and value-free. Gould, however, treats the history of science like political history, with which his readers are more familiar: a history in which human motives and errors from the past will inevitably recur. He thus skillfully promotes a doubt that the biological roots of human behavior can ever be explored scientifically.

Politicizing and Publicizing Science

A left-wing group called "Science for the People," of which Gould is a member, has been particularly active in campaigning against such studies. Instead of focusing, in the earlier tradition of radical groups, on defects in our political and economic system that demand radical change, this group has aimed at politicizing science, attacking in particular any aspect of genetics that may have social implications. Their targets have included genetic engineering, research on the effects of an XYY set of chromosomes, sociobiology, and efforts to measure the heritability of intelligence. Several years ago Gould co-signed their intemperate attack on E. O. Wilson's *Sociobiology: The New Synthesis.*[5] Now, in *The Mismeasure of Man,* he has extended the attack to cognitive psychology and educational testing, because they may reveal genetic differences.

Gould has spelled out explicitly his ideological commitment, and also its influence on his science. Thus, as we shall see, his main scientific contribution has been the claim that evolution has occurrred mainly through revolutionary jumps, rather than by small steps. Both in a "Dialectics Workshop"[6] and in a scientific paper[7] he supports this claim with a citation from Marx: "Darwin's gradualism was part of the cultural context, not of nature." He adds that "alternate [sic] conceptions of change have respectable pedigrees in philosophy. Hegel's dialectical laws, translated into a materialist context . . . are explicitly punctuational, as befits a theory of revolutionary transformation in human society." And, "it may also not be irrelevant to our personal preferences [about evolutionary mechanisms] that one of us learned his Marxism, literally at his Daddy's knee." To most scientists (other than those tethered to a party line) such a claim of support from (or for) Hegel is silly, and such an insertion of an ideological preference, whether from the left or the right, is a corruption of science.

These quotations may help us to understand why *The Mismeasure of Man* ends up as a sophisticated piece of political propaganda, rather than as a balanced scientific analysis. Gould is entitled, of course, to whatever political views he wishes. But the reader is also entitled to be aware of his agenda.

It may also be pertinent to comment briefly on Gould's scientific writing. His claim to have disproved the widely accepted, "gradualist" view of evolution has had great appeal for science reporters, but it has been subject to intense criticism by his professional colleagues. Of course, controversies in science are not rare, and it would not be appropriate here to try to judge Gould's stature as a scientist. It is pertinent, however, to note features of his professional writing remarkably similar to those that I have criticized in *The Mismeasure of Man.* In both contexts he focuses primarily on older approaches to problems in which genetics is now central; he picks his history; and he handles key concepts in an ambiguous manner. Moreover, he is fond of artificial dichotomies that oversimplify complex issues: evolution by leaps versus evolution by gradual steps; biological determinists versus environmentalists; general intelligence versus specialized intelligence.

While Gould has made a valuable scientific contribution in providing evidence that marked fluctuations in rate are common in evolution, the most general professional criticism is that in dramatizing this contribution he has set up a nonexistent conflict with the prevailing gradualist view. For he proceeds as though gradualism implies a relatively constant rate as well as small steps. But even Darwin recognized that the rate of evolution might vary widely, and modern investigators have demonstrated many mechanisms that contribute to such fluctuation.

Neo-Lysenkoism

In *The Mismeasure of Man* Gould fails to live up to the trust engendered by his credentials. His historical account is highly selective; he asserts the nonobjectivity of science so that he can test for scientific truth, flagrantly, by the standards of his own social and political convictions; and by linking his critique to the quest for fairness and justice, he exploits the generous instincts of his readers. Moreover, while he is admired as a clear writer, in the sense of effective communication, he is not clear in the deeper sense of analyzing ideas sharply and with logical rigor, as we have a right to expect of a disciplined scientist.

It has been uncomfortable to dissect a colleague's book and his background so critically. But I have felt obliged to do so because Gould's public influence, well earned for his popular writing on less political questions, is being put to mischievous political use in this book. Moreover, its success undermines the ideal of objectivity in scientific expositions, and also reflects a chronic problem of literary publications. My task has been all the more unpleasant because I do not doubt Gould's sincerity in seeking a more just and generous world, and I thoroughly share his conviction that racism remains one of the greatest obstacles.

Unfortunately, the approach that Gould has used to combat racism has serious defects. Instead of recognizing the value of eliminating bias, his answer is to press for equal and opposite bias, in a virtuous direction—not recognizing the irony and the danger of thus subordinating science to fashions of the day. Moreover, as a student of evolution he might have been expected to build on a profound insight of modern genetics and evolutionary biology: that the human species, and each race within it, possesses a wide range of genetic diversity. But instead of emphasizing the importance of recognizing that diversity, Gould remains locked in combat with a prescientific, typological view of heredity, and this position leads him to oppose studies of behavioral genetics altogether. As the reviewer for *Nature* stated, *The Mismeasure of Man* is "a book which exemplifies its own thesis. It is a masterpiece of propaganda, researched in the service of a point of view rather than written from a fund of knowledge."

In effect, we see here Lysenkoism risen again: an effort to outlaw a field of science because it conflicts with a political dogma. To be sure, the new version is more limited in scope, and it does not use the punitive powers of a totalitarian state, as Trofim Lysenko did in the Soviet Union to suppress all of genetics between 1935 and 1965. But that is not necessary in our system: A chilling atmosphere is quite sufficient to prevent funding agencies, investigators, and graduate students from exploring a taboo area. And such neo-Lysenkoist politicization of science, from both the left and the right, is likely to grow, as biology increasingly affects our lives—probing the secrets of our genes and our brain, reshaping our image of our origins and our nature, and adding new dimensions to our understanding of social behavior. When ideologically committed scientists try to suppress this knowledge they jeopardize a great deal, for without the ideal of objectivity science loses its strength.

Because this feature of science is such a precious asset, the crucial lesson to be drawn from the case of Stephen Jay Gould is the danger of propagating political views under the guise of science. Moreover, this end was furthered, wittingly or not, by the many reviewers whose evaluations were virtually projective tests of their political convictions. For these reviews reflected enormous relief: A voice of scientific authority now assures us that biological diversity does not set serious limits to the goal of equality, and so we will not have to wrestle with the painful problem of refining what we mean by equality.

In scientific journals editors take pains to seek reviewers who can bring true expertise to the evaluation of a book. It is all the more important for editors of literary publications to do likewise, for when a book speaks with scientific authority on a controversial social issue, the innocent lay reader particularly needs protection from propaganda. Science can make a great contribution toward solving our social problems by helping us to base our policies and judgments upon reality, rather than upon wish or conjecture. Because this influence is so powerful it is essential for such contributions to be judged critically, by the standards of science.

NOTES AND REFERENCES

1. S. J. Gould, *The Mismeasure of Man* (New York: Norton, 1981).
2. H. H. Goddard, "Mental Tests and the Immigrant," *Journal of Delinquency* 2 (1917):243.
3. Gould's reference to "enhancing potential" is revealing, for it confuses *genotype* (an inborn range of potential) and *phenotype* (the actual *ability* developed within that range). He should have spoken instead of enhancing performance, or of enhancing the development of potential. This is not a trivial semantic distinction: It is essential for any clear analysis of the interaction of genes and environment. Gould's language suggests that he either does not fully understand, or feels compelled to ignore, this key concept in genetics.
4. Gould's broad generalization ignores the fact that the disadvantaged Chinese and Japanese in this country have consistently scored even higher than Caucasians. Moreover, in including sex discrimination in the IQ controversy, he is straying far from reality. In fact, females average the same as males on standard IQ tests: They perform slightly better on verbal tests, and slightly worse on spatial tests, but the tests are constructed to balance these differences.
5. E. Allen et al., Letter, *New York Review of Books* (November 13, 1975):43. See also "Sociobiology Study Group of Science for the People" in *BioScience* 26 (1976):182. This article includes the remarkable statement that "We know of no relevant constraint placed on social processes by human biology."
6. S. J. Gould, "The Episodic Nature of Change versus the Dogma of Gradualism," *Science and Nature* 2 (1979):5.
7. S. J. Gould and N. Eldridge, "Punctuated Equilibria: The Tempo and Mode of Evolution Reconsidered," *Paleobiology* 3 (1977):115.

15

The *Mismeasure of Man* Controversy

When Stephen Jay Gould published a rebuttal of my sharp criticism of his book The Public Interest *invited me to write a reply, which follows. Though it may seem unfair to reprint only one side of such an exchange, I hope the article will be read as an effort to clarify further my views on some of the main issues of this book, rather than as a further criticism of Gould. The interested reader will find his forceful statement on page 148 of the journal.*

Since Gould had published a vigorous denunciation of Lysenkoism—a politicization of science that killed all genetics in the Soviet Union for twenty-five years—some colleagues felt that my use of the term neo-Lysenkoism was too harsh. For they, along with Gould, see the essential feature of Lysenkoism as proscription of a field of science by the state—an aim that I would never ascribe to him. My defense is that I defined neo-Lysenkoism carefully, in terms of purpose ("an effort to outlaw a field because it conflicts with a political dogma") rather than in terms of methods. As I pointed out in the original article, in our political system the punitive powers of a totalitarian state are not needed to outlaw a field: a chilling atmosphere is quite sufficient. Indeed, as we shall see later, this is precisely how research on the effects of an extra X or Y chromosome has been brought to a halt in this country. I therefore believe that the term neo-Lysenkoism is a useful one for sharpening our focus on the ideological aims of a movement, especially since that movement has considerable sentimental appeal.

The Public Interest (Spring 1984):152.

As my article stated, I do not doubt Stephen Jay Gould's sincerity in seeking a more just and generous world. Moreover, since he considers the article unfair and cruel, he deserves credit for confining his reply to the intellectual issues and for maintaining civility in our personal contacts. But I do not think I was cruel, though I certainly was severe. In trying to treat such serious issues responsibly I could not afford to mute any painful facts or inferences; and Gould's own colorful style does not invite gentleness. Readers of his columns and books will recall how vigorously he has questioned the integrity and the competence of those whose views offend him (see, for example, his "Of Crime, Cause, and Correlation" in *Discover*, December 1983).

Gould charges that I have built on a false premise. Tabulating a complete list of reviews, he notes that the majority of those by psychologists and "distinguished scientists in related fields" is positive. But numbers alone, without evaluation of qualifications, carry little weight: One would not pay much attention to a family doctor reviewing a book on the genetics of schizophrenia. I therefore felt justified in focusing on major journals, where, as I claimed, scientists *close to the field* were negative.

Psychologists are a diverse group, and most would not be expert enough in behavioral genetics or psychometric theory to judge critically the heart of Gould's book; his novel arguments about general intelligence and heritability. Specifically, he identifies three favorable reviewers professionally, and it is interesting that they are an educational psychologist, a professor at a private psychiatric institution, and a sociobiologist.

Gould further charges that my grounds for dismissing four laudatory reviews by scientists were dubious. Three of these speak for themselves, as cited in my article. In the case of the fourth, *Scientific American*, my claim of a blind spot in that excellent journal might seem ad hoc, so I will provide an example. In 1974 Leon Kamin, in *The Science and Politics of IQ*, claimed to find a fatal flaw in every published study of the heritability of IQ, and he concluded that the scientific method requires us to accept the "null hypothesis" that the magnitude is zero. This is nonsense, for zero is a specific value, and in the absence of proof it is no more likely than any other value. Depending on how severely one views the flaws in the evidence, the scientific answer should be either that we are prepared to accept an imprecise, broad range of values, or that we do not know. *Scientific American* chose as its guest reviewer an astronomer known to be sympathetic to the extreme environmentalist view, and he supported the pseudoscientific argument for the null hypothesis.

I realized that citing Jensen's review would turn away some readers; and it was obvious that this review, as Gould suggests, might be biased in my favor. Moreover, in my opinion Jensen has accepted estimates of the heritability of IQ that are almost certainly too high. Nevertheless, since I found his technical analysis of Gould's book highly competent and comprehensive, I would betray a scholarly obligation if I ignored such a major review in order to avoid criticism. The interested reader may wish to compare the tone of this article, in the *Contemporary Education Review*, with that of Richard Lewontin in the *New York Review of Books*, or the *Scientific American* review noted above (July 1975).

Moving on to my own analysis of the book, Gould charges me with falsely impugning his scholarship, by ignoring his recognition that Goddard's population sample, in his famous Ellis Island study, was not random. But while he did indeed discuss Goddard's sampling, he then made a huge extrapolation: "Could anyone be made to believe that four-fifths of any *nation* were morons? [emphasis added]." Compare this summary by Gould with Goddard's first sentence, which I cited: "This is not a study of immigration in general but of six small highly selected groups."

Actually, Goddard's research was so primitive that there is little point in arguing over his sampling, and over whether his effort to balance a bias in selection reflected prejudice or scientific naivete. Gould ignores something much more fundamental: a quotation from Goddard, which I cited, expressing confidence that the children of these immigrant "morons" will be "of average intelligence . . . and good citizens." This citation contrasts remarkably with Gould's heading for this story: "Preventing the immigration and propagation of morons."

I am surprised that Gould would bother with my citation from a review that discussed the 1917 Army "Alpha" tests, since I referred to these tests only briefly, as "incredibly crude." No doubt public knowledge of the tests did play a role in the Immigration Restriction Act of 1924 (which I, also from an immigrant background, deplore as much as Gould). However, since he made the further strong claim that the act "clearly reflected the lobbying of scientists and eugenicists," it seemed fair for me to note the reviewer's statement that no psychologist testified before Congress, and that the reports of the House Committee do not mention intelligence tests. Far more to the point, Gould does not reply to my main criticism of the historical part of his book: his implication that the horror stories of the past are a permanent paradigm for the field.

In commenting on my critiques of his arguments on general intelligence and on heritability, Gould emphasizes his separation of the claim for the meaning of IQ and the claim for its heritability. But when he speaks of "the only theoretical justification that hereditarian theories of IQ ever had," the separation becomes fuzzy. The phrase "hereditarian theories" has no clear meaning, and it is understandable that I interpreted it as a pejorative synonym for the study of heritability (the fraction of the variance in intelligence that is due to inheritance). If Gould did realize that his denunciation of IQ as an artificial composite had no implication for the heritability of its individual factors, it is strange that he should have avoided mentioning that fact, and avoided even defining heritability. Indeed, in his comment he uses "heritability" and "inheritance" as synonyms, thus continuing to treat the issue in terms of the outmoded artificial dichotomy of nature versus nurture.

On this topic, Gould does not discuss my substantive objections to his reification and factoring arguments against the concept of general intelligence, and my suggestion that its measurement can still have practical value (if the limitations are recognized), even as more refined analyses, differentiating various kinds of intelligence, are being developed. I would contrast his assault on IQ with Howard Gardner's recent constructive plea in *Frames of Mind* (Basic Books, 1983), also written for a broad audience, for recognizing multiple intelligences.

I was of two minds about discussing Gould's science, and I will agree that the resulting exposition was too brief to convey a clear picture of "punctuated equilibrium." But his picture is not clear. In recent papers he has so attenuated the concept that I no longer disagree—but I also find little novelty. Meanwhile, in the press he continues his bold claims for a revolutionary reinterpretation of Darwinism, and maintains the image of David slaying the Goliath of the evolutionary establishment.

Since Gould has characterized my criticism of his science as "ugly," let me defend its objectivity by noting that his work has elicited a remarkably large number of opposing articles by leading evolutionists, criticizing in particular his way of formulating problems and his claims for originality. In one example, Ernst Mayr, writing in the *American Naturalist*, has politely but firmly torn apart the caricature of neo-Darwinian theory as, "panglossian paradigm," that Gould and his co-author Lewontin attack. And a number of letters about my article, from members of the National Academy of Sciences in Gould's field, have emphasized that such a sharp criticism of his science and its political bias was long overdue. It is clear that Gould, however effective as a teacher and a popularizer, is not a scientist's scientist.

Finally, neo-Lysenkoism: I can understand Gould's resentment at this charge, which would indeed be unjustified if it implied any of the definitions that he discusses. But I defined the term carefully: "an effort to outlaw a field of science because it conflicts with a political dogma." And the Lysenkoist model comes even closer because Gould is dealing with a branch of genetics. While I do not doubt that he would be opposed to proscription by government fiat, and I know he has written a vigorous attack on the original Lysenkoism, I cannot ignore the fact that he is affiliated with Science for the People, and his book supports their campaign against behavioral genetics—a campaign that has, for example, halted research in this country on the effects of an extra Y chromosome.

Gould ends by asserting his right, as a scholar, to expose the biases of Western thought. But in *The Mismeasure of Man* he is not an innocent scholar exposing novel arguments to critical evaluation by other scholars. He is a public figure, using his scientific credentials to influence the attitude of a much wider lay public toward mental testing and behavioral genetics. And as a public figure he is not immune to unvarnished criticism. One could reasonably charge that my attack was too blunt: That is a matter of taste and judgment. But I do not accept his characterization of my explicit exposition as invective and innuendo: Indeed, it is surely the opposite of innuendo.

To end on a more positive note: I think Stephen Jay Gould is a natural treasure as an effective conveyer of science to the public. But I am sorry he handled the IQ controversy and related topics the way he did in *The Mismeasure of Man*—sacrificing scientific integrity to hyperbole for political purposes.

16

Review of *Not in Our Genes*

I have already discussed this book in a letter in Part One (item 5). And it continues to amaze me that such intelligent scientists as these three authors can persist in a sort of religious conviction that dialectical materialism is the key to success in science: for the evidence to the contrary is striking. The explanation for their faith may be their continued admiration for another aspect of Marxism, which has nothing to do with science: its idealistic stated aims. But while many of us who shared that sympathy and hope in the 1930s have lost confidence in the translation of this dream into practice, I do not object if the authors have political beliefs quite different from mine. But I cannot understand why they should apply this ideology so forcefully to a critique of something that is much more objective: the phenomenally successful scientific enterprise.

The most interesting feature of the book is the marked shift from an earlier position (which they no longer acknowledge): denial of the importance of genes for behavioral differences. The authors now snatch from their opponents the position that both genes and environment, interacting, are important. But one must wonder whether this shift was not merely one of strategy, without a real conversion: for the book retains a title that simply does not fit this new thesis.

The book also attacks "reductionism," in ways that are hard to relate to the usual meaning of the term. This term was discussed in more detail in Part Two, where I addressed "Sociobiology, Human Individuality, and Religion."

Commentary (January 1985):71. Review of a work by R. C. Lewontin, S. Rose, and L. J. Kamin (New York: Pantheon Books, 1974).

For over a decade a group of scientists on the radical Left, Science for the People (SFP), has pursued a campaign against studies of human behavioral genetics. *Not in Our Genes* is a major contribution to that campaign, and the authors make no secret of their political purpose:

> We share a commitment to the prospect of the creation of a more socially just—a socialist—society. And we recognize that a critical science is an integral part of the struggle to create that society, just as we also believe that the social function of much of today's science is to hinder the creation of that society by acting to preserve the interests of the dominant class, gender, and race.

Given this ideological commitment, it is not surprising that the authors condemn human behavioral genetics, intelligence testing, and sociobiology as worthless and even pernicious. Moreover, they attack the motivation as well as the conclusions of scientists engaged in these fields, projecting upon them an aim as frankly political as their own, but opposite in intention. All work in these areas is seen as serving a prejudiced society: "It is precisely to meet the need for self-justification and to prevent social disorder that the ideology of biological determinism has been developed." Even more, those whom they label "biological determinists" are accused of continuing the earlier misuses of genetic theories to rationalize racial discrimination.

In their passion to discredit genetic studies, the authors even go so far as to reject Seymour Kety's classic demonstration of the major role played by heredity in schizophrenia. Yet studies such as Kety's are in fact a necessary step toward the identification of abnormal genes; and while Lewontin et al. clearly fear that this knowledge will be used for reactionary eugenic measures, in actuality it provides the most promising approach, through the fantastic power of molecular genetics, to understanding the disease and to developing specific therapies.

Some of the views of the authors on behavioral genetics and on psychometric testing seem idiosyncratic or just plain silly: for example, in psychometric testing "human action is itself reduced to individual reified lumps objectified in the black box of the head." And in a gross distortion of history and of science the authors write:

> [IQ] test items that differentiated boys from girls . . . were removed, since the tests were not meant to make that distinction; differences between social classes, or between ethnic groups or races, however, have not been massaged away, precisely because it is these differences that the tests are *meant* to measure.

This statement is inexcusable. As anyone in the field knows, boys average slightly better on spatial items and girls slightly better on verbal items, and their overall averages are equalized by adjusting the proportions of various kinds of items, not by removing differentiating ones. The other groups mentioned cannot be equalized in this way because they do not have compensatory areas of high performance, even in nonverbal tests.

While most of *Not in Our Genes* repeats earlier arguments made by Science for the People, it is important to note one novel position that seems to represent a major shift. After many years of denouncing so-called "biological determinists" and struggling to convince the world that genes have little to do with behavioral differences between individuals, Lewontin and his colleagues (and also Stephen Jay Gould in his recent writings) now state their conviction that intelligence is the product of interactions between genetic potentials and environmental inputs, and that both sources are substantial. This, as it happens, is the position that has long been held by serious students of the subject. Unfortunately, although the authors now have an opportunity to end a lot of sterile polemics, they deny that their opponents hold this interactionist view; the denigrating epithet "biological determinist" is repeated by them, in a depressingly familiar political tactic, on virtually every page. Moreover, the title of their book forces one to suspect that the authors' own conversion to interactionism is a matter less of conviction than of tactics. One's skepticism is reinforced by their comment on J. B. S. Haldane and H. J. Muller, two distinguished earlier geneticists and Marxists who "argued (along lines that we would not) that important aspects of human behavior were influenced by genes."

In addition to attacking what they call "biological determinism," the authors also spend a good bit of time tilting at "reductionism": the attempt to understand complex systems in terms of the properties of their components. They condemn contemporary pharmacological research as simplistic because it seeks drugs that hit a specific target; and they even object to testing the blood-alcohol level of possibly drunken drivers because a variety of different moods can be associated with the same level. In place of this "reductionist" approach they expound their faith that dialectical materialism offers a more powerful analytic tool:

> We would counterpose the understanding of . . . revolutionary practitioners and theorists like Mao Tse-tung on the power of human consciousness in both interpreting and changing the world, a power based on an understanding of the essential dialectic unity of the biological and the social . . . as ontologically coterminous.

But if the insights of the dialectic are so valuable for biomedical science, one wonders why the Soviet Union has not developed a single useful antibiotic. It is sad indeed to see how these intelligent scientists, in their dedication to seeking a better society through Marxism, feel obligated as well to apply the murky principles of dialectical materialism to science.

Why trouble to review such doctrinaire material? The reason is that some highly respected publications have greeted this book enthusiastically, as a serious scientific contribution. Yet as I hope to have made clear, however briefly, the fundamental aims of *Not in Our Genes* lead to a distorted picture, feed a growing anti-science sentiment, and undermine the very foundation of science: the commitment to objectivity. To be sure, scientists approach the ideal of objectivity only imperfectly, and sometimes are guilty of unconscious bias; but if their practice remains honest, the resulting errors can eventually be corrected. By contrast, any deliberate introduction of ideological preconceptions, à la Lewontin, Rose, and Kamin, compromises the integrity of science, today no less than in the time of Galileo.

17

XYY: The Dangers of Regulating Research by Adverse Publicity

Science for the People has not been very successful in convincing people that genes are unimportant for human behavior. In consequence, Lewontin (as the preceding piece shows) and Gould are now claiming the position long held by those whom they mislabel as biological determinists: that neither genes nor environment, but interactions between them, are the source of the differences in individuals' abilities. However, at an earlier time, pressing their claim that it was false to believe that genetic studies could help us to understand human behavior, the organization was successful in suppressing an ongoing research project at Harvard University. This longitudinal study, starting at birth, was designed to identify the consequences (still not well understood) of having an extra X or Y chromosome. This successful suppression clearly has discouraged other investigators from undertaking similar studies.

In this article I try to analyze the issues, both academic and scientific, and also to explain why this campaign, drawing on understandable public concern over stigmatization of children, was so successful.

In the 1960s advances in cell biology made it easy to identify people who have abnormal numbers of chromosomes. The behavioral and physical consequences are already well known for some such chromosomal abnormalities but remain obscure for others. Accordingly, six years ago Dr. Stanley Walzer (a psychiatrist) and Dr. Park Gerald (a geneticist) set up at Harvard Medical School a long-term project to detect certain

---

Harvard Magazine (October 1976):26–30.

chromosomal aberrations in newborns, and to determine what effects, if any, the aberrations might have on later development. Last May, after nearly a year of adverse publicity for the project and after much personal harassment of Dr. Walzer, the investigators abandoned the testing program.

This development raises several questions. What is the ideological basis for this and other attacks on research in behavioral genetics? What effects are these attacks likely to have on advances in the biomedical sciences? How can the public, increasingly concerned with the ethics of medical research, acquire adequate information and perspective on the balance of risks and benefits? What are the obligations of a university to a faculty member who is virulently attacked by another? Before considering these problems, I shall review the history of the XYY case and try to evaluate the substance of the charges.

The History

The normal pattern of human chromosomes is 23 pairs, of which the sex-determining ones are normally an XX pair in a female and an XY in a male. In the project of Drs. Walzer and Gerald all the male newborns at the Boston Hospital for Women were tested, and a follow-up was offered for those with an extra female chromosome (XXY), for those with an extra male chromosome (XYY), and for a control group.

About 45 XXY and XYY individuals were detected among the 15,000 baby boys tested. The parents of these infants were told that their child had an extra chromosome, and that its possible effects on his health or behavior are not yet known. The follow-up that was offered involved several visits a year by the psychiatrist and also visits by an additional trained observer. This close follow-up has a dual purpose: to compare the child's behavioral development with the norms for the general population, and to try to help with any problems that might arise (using the supporting facilities of the hospital if necessary). The study is thus not experimental: it is rather in the ancient clinical tradition of trying to detect, observe, and ameliorate a potential health problem in which research is still at the natural-history—observation—stage. The only novel feature is the use of modern laboratory methods to reveal the presence of the abnormality.

The study was unfortunately complicated by the development of a widespread public misconception. Shortly before it was begun, investigators in several countries found that XYY was ten to twenty times more frequent in inmates of institutions for the criminally insane than in the general male population (1-2 percent versus 0.1 percent). This finding clearly means that an extra Y chromosome increases an individual's probability of incarceration in such institutions. However, the probability is still very small. Unfortunately, sensational popular articles erroneously reported that an extra male chromosome uniformly causes excessive aggressiveness, and they spoke of discovery of "the gene for criminality."

This misconception could obviously create anxiety and the danger of stigmatization, and the investigators took whatever steps they could to prevent these effects. Only when the parents specifically asked were they told that the extra chromosome was a Y, and Dr. Walzer then explained in detail that we lack clear knowledge about the significance of the aberration, contrary to the public impression. Moreover, extreme precautions were taken to maintain privacy, and the records were locked up in the psychiatrist's office rather than in the hospital files. Nevertheless, a year ago this quiet, discreet study was denounced by a small group of political activists (Science for the People), led by microbial geneticists Jonathan Beckwith of Harvard Medical School and Jonathan King of M.I.T.

The critics began with the customary procedure for an academic dispute, filing a formal complaint with the medical school's administration. The committee to which it was referred moved slowly, and after its report the administration found that an additional committee had to pass judgment. Meanwhile, Dr. Beckwith and his associates decided to hold a press conference and public meetings. The meetings were not debates in the academic tradition: they were rallies launching a crusade against evil. The public debate over this issue had some unfortunate immediate effects. During the next month, Dr. Walzer and his family received about twenty anonymous abusive, and sometimes even threatening, phone calls.

The medical school's final review committee rejected the complaint, concluding that the study had been properly evaluated and approved by the several appropriate hospital and school committees, and that the conduct of the investigators did not violate any ethical principles. Dr. Beckwith appealed to the faculty, and his resolution was defeated 199 to 35.

For several months after the Harvard Medical faculty vote, there was no further attack, and it seemed that the matter had been settled. But the alleged assault on XYY children by Dr. Walzer had come to the attention of the Children's Defense Fund, an organization of public-interest lawyers concerned with the victimization of children. A representative visited Dr. Walzer to express his group's interest in terminating the study.

Faced with the possibility of an even broader campaign of public criticism, Dr. Walzer decided to give up further screening (but to continue with the patients already under study). To be sure, screening had been scheduled to end within less than a year. Nonetheless, its abandonment in response to pressure will surely influence the willingness of future investigators to undertake similar studies.

The Charges

Let us now analyze the charges. The main objection has been that XYY identification will socially stigmatize a child. In view of current public misinformation, this charge would indeed be justified if the identity of the child were made public. But as we have seen, the physician-investigators have taken careful precautions against that possibility, and with complete success.

A closely related concern was that the children would be harmed by the self-fulfilling prophecy that they would develop behavioral abnormalities. Self-fulfilling prophecy is indeed found in many social interactions, including some of those that occur between parents and children, but it varies widely in degree from one kind of behavior to another. It is thus not a solid scientific law, with firm predictive value—otherwise, we could make all children intelligent and well-behaved just by wishing hard enough. Indeed, in the XYY project in particular the fear of the self-fulfilling prophecy has proved unfounded, because Dr. Walzer has not in fact observed any of the excessive aggressiveness that would be predicted.

The critics further argue that the study has no scientific value because the only way to correct for the possible self-fulfilling prophecy would be to falsely label a control normal group as XYY—which would be immoral. I am somewhat surprised that this criticism has been widely accepted among professionals, for I believe the argument is based on an excessively narrow conception of the scientific method. In dealing with human subjects an ethical investigator can rarely set up the ideal controls; but even without ideal controls, one can detect effects when they are large enough. XYY would have little practical importance if its effects in the present study should turn out to be too small to be established by comparison with the range of ordinary clinical experience. However, if a substantial deviation from the normal range should appear repeatedly (even if not uniformly) in a series of XYY subjects, then the extra Y chromosome could reasonably be held responsible. The physician must be guided by the most *probable explanation*, rather than wait for absolute certainty.

In defending the scientific aspects of this study, I do not suggest that it could not have been improved. The scientific literature is filled with disputes about how various pieces of research might have been better designed. But these problems are irrelevant to questions of ethics—unless a study is so utterly valueless that it has no benefit to balance against even minimal risk.

A criticism more closely related to ethical problems was that the investigators had assumed parental consent for the chromosomal test, or had obtained it too casually. The investigators and the review committees accepted this charge and developed an elaborate consent form (indeed, probably too elaborate to be useful for most patients). They also improved the procedure by requesting consent at an earlier time, before the mother was in labor.

Nevertheless, one might even question whether it is self-evident that a test for a chromosomal abnormality should require informed consent. Although this requirement has now been generally accepted for experimental procedures that create *risk of physical harm,* what we are dealing with here is the gathering of prognostic information that is disturbing (i.e., information with possibly unpleasant implications for the future), which leads to *anxiety* rather than *risk.* A physician seeks and conveys such information every day. Would it be appropriate to require informed consent for performing a test for albumin in the urine, which one may also not be able to cure?

It has been suggested that this problem could be solved if the investigators were to identify the XYY babies but not inform the parents. This attractive solution unfortunately fails on several grounds: (1) Continued cooperation of the families for many years could not be expected without the investigators' supplying a good reason; (2) active concealment of the finding would be illegal, since in Massachusetts patients now have access to their medical records; (3) even without this law, a physician would be liable for malpractice if he had failed to provide a warning that could conceivably have prevented a later problem; (4) if a physician withheld such information, he could justifiably be accused of elitism and paternalism—with Science for the People perhaps casting the first stone.

The critics have also suggested that the moral problem could be eliminated by screening adults rather than infants. But experts agree that only a long-range, "prospective" study, starting early in life, can provide a clear picture of the effects of the extra chromosome on development. Moreover, such a study can tap a representative, unbiased population sample, which is difficult to obtain with adults. (For example, XYY might also predispose to socially useful forms of aggressiveness; but one cannot obtain random adult samples for testing this possibility.) Also, some chromosomal abnormalities predispose a person to infections that occasionally prevent survival into adult life, and so a study of adults might underestimate the seriousness of the condition. Finally, there is a humanitarian consideration as well: from what we know about other chromosomal abnormalities, we can expect an XYY person to have an increased likelihood of developing problems, and it is obviously desirable to try to help as early as possible.

I conclude that the case for condemning the XYY study is far from overwhelming. The questions raised would be thoroughly legitimate for discussion in a committee or a classroom concerned with medical ethics, but they surely do not justify a public hue and cry. Moreover, the failure to convince the faculty was hardly surprising, since the critics built their case on serious misconceptions about clinical research, medical practice, self-fulfilling prophecy, and human genetics. Finally, it is ironic that the critics are opposing the very research necessary to correct the public misconception about the significance of XYY.

The Ideological Purpose

Why, then, should an activist political group make such an issue of XYY, with all its scientific and ethical ambiguities and with so few "victims" involved, when much larger and more indefensible medical problems abound?

The key, I would suggest, is not primarily concern for the innocent children, though that is surely present. It is the conviction that any attention to genetic factors in behavior will have reactionary social consequences, just as Social Darwinism and the eugenics movements of the nineteenth century did. In a letter published last year, members of Science for the People stated that attention to genetic factors in behavior "only serves to propagate the damaging mythology of the genetic origins of 'antisocial' behavior," and so it interferes with the job of eliminating the social and economic factors involved in such behavior.

In actuality, however, modern genetics does not lead at all to the same conclusions as the false biological inferences and crude social analogies of the Social Darwinists. We now know that evolution selects for cooperative as well as for competitive tendencies; we also know that behavior is governed neither by genes nor by environment, but by their interaction. Accordingly, the development of behavioral genetics, helping us to recognize individual differences in genetic potentials, drives, and patterns of response, should also help us to equalize opportunities for maximal individual self-fulfillment.

The XYY story also raises the old problem of how a democracy can protect itself from those who would use freedom of speech to deprive others of significant freedoms—in this case, freedom of inquiry, and the patient's freedom to know. It is understandable that an advocacy group should feel entitled to resort to public appeal in fighting what it perceives as evil, and it is clear that public airing of such horror stories as the Tuskegee study (involving the deliberate nontreatment of known syphilitics) has helped to identify serious problems. Unfortunately, in an ambiguous case, like the present one, a crusade against evil results in excessive polarization, which distorts the analysis of the issues and impairs the prospects for free intellectual exchange between academic colleagues. It also may create an irrational public response to a demagogic appeal.

Academic Responsibility

Academic institutions faced with this problem have usually stood silently on their dignity, hoping that the public would soon recognize a weak case. But in the current atmosphere, with growing public involvement in the regulation of research, perhaps institutions should recognize their vulnerability and take more initiative to ensure that the public has access to information on all sides of an issue. In addition, one might question whether academic freedom includes the right of a faculty member to carry an academic controversy to the public before it has been adequately explored within the institution, and the right to present an issue of policy in a way that publicly impugns a colleague's reputation. It seems particularly unfair for an individual to bear the brunt of such an attack, when the institution supports his activity. Yet Harvard Medical School made no effort to protect Dr. Walzer.

Perhaps our institutions, while tolerating dissent, might set limits on its style. For we must recognize that we are dealing not simply with legitimate dissent. Just as Lysenko destroyed all of genetics in the Soviet Union from 1935 to 1969, Science for the People aims to destroy the field of human behavioral genetics. And we would be naive not to recognize that an opposition to certain ideas underlies its attack on allegedly harmful research activities.

It is particularly disturbing that even though the case against the research was weak and was overwhelmingly rejected by a responsible faculty, it evidently elicited a favorable reaction from a good deal of the public—even including people who did not share the general political convictions of the critics or approve of their methods, and who knew that the faculty overwhelmingly supported the study. Moreover, the lawyers who delivered the *coup de grâce* undoubtedly sensed a sympathetic public response.

Since we now face proposals, such as those of Senator Edward Kennedy, for increased public participation in the regulation of research, it seems important to try to understand the public sympathy for the XYY attack. In part it was no doubt based on the current general loss of confidence in authority, and hence in the value of professional credentials. Moreover, resentment over the unsatisfactory economics and distribution of medical care has contributed to suspicion of the quite different area of medical investigation. These attitudes helped create a David-and-Goliath image of the critics and the medical faculty. But the most important factor was probably the strong emotional impact of the idea of stigmatizing innocent children.

This impact must give us pause, for it reflects a major difference between laymen and professionals in their perspective on the ethics of research. The layman finds the risks of harm from research highly visible, while the consequences of ignorance are likely to be seen as acts of God. The professional, however, is acutely aware of the doctor's continuing responsibility for dealing with problems that are not yet, but might become, understood; hence, sins of omission loom as large as sins of commission. The layman also is inclined to take for granted the benefits of present knowledge, while the professional knows that we had to take risks to acquire that knowledge and we will have to take risks to expand it. Accordingly, it seems essential that any public involvement in regulating research be set up in a highly responsible, well-informed manner, avoiding the emotional pitfalls of direct public appeal. Otherwise, valuable advances could be paralyzed, on a large scale, by demands for absolute freedom from risk.

18

Concerning Human Behavioral Genetics

Since my defense of the study of human behavioral genetics has sometimes been accused of racism it seems appropriate to include an early published letter that should help to clarify my views.

The history behind this letter is pertinent. At several successive annual meetings of the National Academy of Sciences in the late 1960s William Shockley pressed for a study that would try to settle the question of the heritability of the observed racial differences in IQ. A committee set up by the academy rejected this demand but defended human behavioral genetics as a legitimate field of inquiry. Not having been a member of that committee, I wrote this letter, published in a newsletter distributed to members, in order to endorse the recommendation of the committee and to try to clarify further some of the issues.

On rereading the letter I still find the arguments sound, particularly the argument that any statistical differences in abilities between races should be irrelevant if we define racial justice in terms of eliminating discrimination against individuals. In contrast, if we define it in terms of proportional representation of races in all occupations we are building on the underlying assumption of equal distribution of talents, and it then becomes relevant to test that assumption. Obviously, I greatly overestimated the possibility that such an effort at carefully dissecting the logic of the problem could influence a fundamentally political process.

Dear Mr. President:

I should like to support the suggestion of the *ad hoc* Committee on Genetic Quality, chaired by Kingsley Davis, that it would be highly

---

National Academy of Sciences, *President's Letter to Members* (July 1970):17.

desirable for the Academy to encourage the study of human behavioral genetics (psychogenetics). For though the application of such knowledge to various social problems is rapidly becoming more urgent, there is widespread resistance to the extension of science into this area. Our difficulties in overcoming racism increase the obstacles, primarily because of failure to distinguish the problem of inheritable differences between races from that of race discrimination.

Racial Differences. It has been traditional for liberals to insist that there are only superficial inheritable differences between races. But I fear that this assumption provides a perilously slender foundation for combating racism. Every geneticist knows that when two populations within a species have been separated and exposed to different selective pressures for many generations, they inevitably accumulate many hereditary differences. Behavioral traits in man can hardly be an exception; and even though their genetic component cannot yet be dissected with precision from environmental influences, it is undoubtedly substantial. Hence human races surely differ, to some degree, in their distribution of genetic potential in various areas of performance, some being statistically stronger in one and others in another. This conclusion, however, does not conflict with the democratic ideal of trying to provide maximal opportunity for self-fulfillment to each individual, within the limits of his capacities and without regard to his race or social background. In such a system any statistical difference between races in the distribution of abilities would be irrelevant: the aim is not to insure identical racial distributions of socioeconomic status, but to remove the discriminatory measures that have artificially created gross inequities.

If inheritable racial differences are thus irrelevant to our society, one must wonder why Dr. Shockley believes it is urgent to search for them, especially at a time when we are struggling to free ourselves from the ugly legacy of slavery. To be sure, if we should deviate from the democratic ideal by adopting a policy of proportional racial representation in all occupations, on the assumption of equal distribution of talents, it would become important to test that assumption. But though pressure has recently developed for such positive discrimination, it seems likely to disappear if we can eliminate our enormous burden of negative discrimination.

Problem of Racism. Our problem, then, is not racial differences, which are a scientifically testable question of fact, but racism, which is a political and social value judgment. This is the view that a person's origins, and not simply his individual capacities, qualities, and achievements, should enter into his evaluation. This doctrine cannot be logically derived from, or refuted by, any evidence for or against racial differences. And racism, by this definition, is highly visible in Dr. Shockley's suggestion, in material distributed to members of the Academy this spring, that a recently observed minor effect in the population genetics of Drosophila might be applicable to man. In this work parents with identical values for a readily quantifiable, polygenic, physical trait were drawn from two populations, whose distribution curves for that trait showed a considerable difference of their means. The interesting finding was that the progeny derived from each set of parents had a mean value shifted slightly away from that of their parents, and toward that of their ancestral stock. If this argument were extended to man one might regard a Negro with a given IQ as slightly inferior, genetically, to a Caucasian with the same IQ. I am amazed that anyone would consider such a second-order effect detectable in the extraordinarily poorly quantifiable field of psychogenetics, even under ideal conditions of environmental parity—let alone with two groups differing enormously in their experience and motivation. And I am horrified by the thought that one might consider such information (if obtainable) a useful guide to social policy; such guilt by genetic association would be profoundly incompatible with the democratic ideal of evaluating a person on his merits.

Encouragement of Research in Psychogenetics. In rejecting Dr. Shockley's campaign, however, we must recognize the danger that an overreaction could discourage the study of human psychogenetics, out of fear that the results might be misused. Such tainting of a scientific field, especially a field that might help enormously in providing a realistic foundation for public policy in several areas, would be most unfortunate. I therefore support the Committee's recommendation that the Academy explore means of encouraging research in psychogenetics. Our educational system would surely become more effective if we could develop more reliable and more refined measurements of innate abilities, limitations, and differences in patterns of learning. In the normal course of events we might simply wait for these improvements to emerge—not only through methodological advances, but also as our society reduces its present gross environmental differences and thus, incidentally, makes possible more accurate genetic measurements. But meanwhile efforts to reach the goal of equal educational performance for pupils at all socioeconomic levels may yield disappointing results, and the hard facts of genetics would then become important, both to avert widespread paranoia and to define more realistic goals. In addition, demographic consequences of the population crisis are also likely to create a need for increased knowledge of psychogenetics. Since these problems are rapidly approaching, and since the required information, and its acceptance by the public, cannot be achieved on short notice, it does not seem too early for the Academy to become concerned.

One particularly appropriate role of the Academy, as an interdisciplinary organization, might be that of encouraging the incorporation of genetics into curricula in departments of psychology and social science and in schools of education. Genetics and evolutionary theory are very recent developments, against the background of thousands of years of recorded prescientific speculation about the nature of man; and they have so far had no significant impact on social policies. The possibility of their making their due contributions to society is surely hindered by the fact that their relevance is still recognized very little in academic departments concerned with such problems. Thus at present hardly any Ph.D. programs in psychology or sociology include a course in genetics; and current "anti-elitist" trends in our culture encourage educators to build on the myth that all children start with equal innate abilities.

In summary: there is strong resistance in our culture to acknowledging that individuals differ widely in their inheritance of intellectual capacities and of other behavioral traits. Yet more accurate identification of these differences should be of great value in trying to provide individuals with optimal opportunities. I therefore hope that the Academy will encourage research in this field and will promote recognition of its relevance to education and to other large areas of social policy. It would indeed be tragic if such studies were inhibited for fear that they might incidentally demonstrate some degree of difference between races in the distribution of inheritable abilities; for any demonstration of such differences should be irrelevant in a society dedicated to providing every individual with the fullest opportunity for developing his capacities.

19

Introduction to *Human Diversity: Its Causes and Social Significance*

I spent the 1973-74 academic year as a Fellow at the Center for the Study of the Behavioral Sciences at Palo Alto, intending to write a book on human diversity. This proved too large an order for my entry into a new field, and the experience yielded instead the background for many of the pieces in this volume.

During that year I organized for the American Academy of Arts and Sciences a series of four weekend seminars at which a number of scholars from different disciplines, and with diverse points of view, discussed the subject. I hoped the evidence from biology would convince the social scientists that genetic factors are important in human diversity, and hence that the current search for equality should emphasize social equality, not the unrealistic goal of equality of achievements. At that time it was a success if one managed to have a civil discussion of this topic among academics of different persuasions; but it was not clear that anyone changed his views very much. We certainly did not have a visible impact on the society around us, when a condensed transcript of the meetings was published several years later; but the problems are still with us, and our society is inevitably becoming more receptive to their open discussion.

The following piece is my introduction to the published transcript.

As our society moves toward greater social justice, we are faced with the challenge of dealing with a wide spectrum of differences among

---

From *Human Diversity: Its Causes and Social Significance*, Bernard D. Davis and Patricia Flaherty, eds. (Cambridge, Mass.: Ballinger, 1976).

individuals and among groups. This diversity is at once an enormous cultural and biological resource and a cause of social and political tension. And though we cannot specify with precision the relative contributions of environmental and hereditary factors to many of these differences, a realistic approach to social policy must recognize the importance of both sets of factors. While legislation and changes in social attitudes can remove many social inequalities—and should be able to eliminate environmental obstacles to individual development—it is not clear to what extent we can hope to modify or to compensate for most hereditary differences.

That the environment has a large role in shaping social and individual diversity is obvious: a person's abilities, drives, and achievements are enormously influenced by cultural traditions, socioeconomic level, education, and life circumstances—including luck. Moreover, social and behavioral scientists have tended in recent decades to rely on environmental factors to explain individual and group differences, partly for methodological reasons and partly because environmental differences appear more susceptible to effective social or political intervention.

The contribution of heredity is less well defined and has received less attention. Behavioral traits are especially difficult to study from this standpoint: they depend on the joint contribution of very many genes; outcomes are the product of the interactions of these genes with the environment and are not determined by genes alone; and we lack direct neurobiological bridges between observations on genes and on behavior. Accordingly, the study of behavior, unlike most other branches of biology and medicine, has benefited little from the recent spectacular success of molecular genetics in probing single genes that have an identifiable molecular product.

Nevertheless, modern developments in evolutionary theory and genetics, resulting in the emergence of the field of population biology, have begun to make constructive contributions to social problems. By focusing on the statistical distributions of genes among populations, and on the role of most genes in defining potentials rather than traits, this new field has provided powerful evidence against earlier typological and deterministic misconceptions of race. More specific contributions of genetics are also proving relevant for public policy. For example, the results of an exhaustive study of schizophrenia, summarized in this volume, illustrate the possibility of demonstrating a large genetic component in a behavioral trait and thus providing a basis for approaching the prevention and treatment of this disease more realistically.

But while it is clear that the identification of hereditary differences could have beneficial uses, it is also clear that this knowledge, like all knowledge, could be used in harmful ways. Because of concern over this possibility—or over the current politicization of the topic—many scientists are reluctant to study such problems. Even apart from these inhibitions, biologists and behavioral scientists rarely communicate with each other about such matters, and they may not even understand each other's terms of reference or language. Moreover, the public debates in this area have been mostly superficial and polemical, rather than informed and analytical. We thus see a large area, with important consequences for public policy, education, and medicine, in which there has been no systematic, critical discussion, and in which neither biological nor social determinism is scientifically justifiable or likely to be helpful.

To promote informed communication on these issues the American Academy of Arts and Sciences organized a series of conferences by a group of scholars from various disciplines and with various—and indeed widely diverging—viewpoints. The original plan was to take up the following topics, in order:

1. A review of evolutionary theory, and particularly its implications for diversity within a species.

2. A review of relevant aspects of genetics, including contributions of molecular genetics to recognition of the breadth of genetic diversity and to understanding of the mechanisms of interaction of genes and environment.

3. The problems involved in measuring and characterizing intelligence and other behavioral traits, and in estimating the contributions of genetic and of environmental factors to individual differences and to group differences in these traits.

4. The implications of current knowledge of human diversity for the psychosocial sciences, including the possibilities for using knowledge of genetic differences in positive ways, and the possible harmful consequences of building policies on false assumptions about the presence or absence of such differences.

5. The question of how public policy can take into account the realities of this biological diversity, given the equally relevant realities of the political process.

We were able to cover the first three topics in considerable depth, but the implications for the social sciences and for public policy were only briefly discussed. It proved impossible to sustain our original intention of avoiding the highly charged subject of specific ethnic differences: after discussing the evolutionary and methodological aspects of diversity the group unanimously requested an opportunity to scrutinize the concrete evidence on variation in IQ, both among socioeconomic and ethnic groups and among individuals. The last meeting was therefore devoted to this topic. In addition, the social scientists did not find it feasible to prepare detailed analyses of the possible implications of the biological evidence for their respective fields. This condensed transcript therefore presents primarily an exposition of the biological background, and the identification of some of the problems.

As was expected, the conference group did not reach a consensus on the origins of group differences. They also differed on the reliability of quantitative assessments of the heritability of intelligence among individuals, but they did agree on the existence of substantial genotypic differences. The exchanges were lively and instructive, and it was particularly gratifying to find that in an appropriate atmosphere persons with widely divergent convictions could discuss the topics dispassionately and with increasing respect for each other's views.

20

Pythagoras, Genetics, and Workers' Rights

While most of the controversies over human genetics have been concerned with its applications to behavior (especially IQ tests and XYY screening), in 1980 the New York Times *ran a series of critical articles on the recent introduction of industrial screening for genetic hypersusceptibility to potentially toxic chemicals. Not surprisingly, some scientists who had been prominent in the XYY debate also opposed such screening, although it was designed to prevent harm to workers. I wrote the following guest editorial in an effort to clarify the issues.*

This controversy did not last very long, in large part because it is thus far possible to detect only a very few rare kinds of hypersusceptibility, and so the practice has limited value. The controversy did, however, result in wide recognition of the importance of certain safeguards: the measurements should be reliable and properly interpreted, and information must be held confidential and not misused. Nevertheless, few would question the principle of using our growing knowledge of this aspect of human diversity to protect individuals from unknowingly taking jobs that are for them particularly dangerous.

Pythagoras, best known for a theorem about the square of the hypotenuse, was also the founder of a widespread religion, Pythagoreanism, that included a rule against eating beans. This rule has long puzzled philosophers but we now have a rational explanation, and it happens to impinge on a current controversy.

Some individuals of Mediterranean origin are poisoned by eating

OpEd Article, *New York Times*, August 14, 1980.

fava beans because these individuals have a genetic defect in a particular enzyme that is known as G6PD. This defect makes red blood cells susceptible to destruction by a component of the fava bean. On the other hand, it also makes these cells resistant to the malarial parasite. Accordingly, the prevalence of the defect in populations from the eastern Mediterranean can be explained by Darwinian selection—that is, when the altered gene arose by a rare mutation thousands of years ago in a malarial region, it rapidly became more frequent in subsequent generations because it allowed persons to survive malaria better and hence to have more progeny.

The distribution of G6PD deficiency has now become a matter of broader interest, with the finding that this enzyme protects not only against the fava bean but also against many potentially toxic chemicals. Tests for this enzyme might therefore be used to detect individuals who would be at exceptionally high risk if exposed to these compounds. Accordingly, some major chemical firms have recently begun to carry out such tests. It is not clear how much testing of this kind is being done, nor is it clear whether the present purpose is primarily to guide employment or to evaluate such methods for the future.

In July, the Institute of Medicine of the National Academy of Sciences held a conference that included this subject within the broader topic of gene-environment interactions. The participants generally agreed that screening for enzyme defects is still a matter for research, and not for industrial practice. However, other critics have objected to the screening in principle, claiming that it shifts the blame for toxic reactions from the employer to the worker, deprives people of the right to free access to a job, and relieves the employer of the obligation to keep down the levels of toxic substances. These objections should be scrutinized carefully, lest we foreclose a potentially valuable activity.

First, the essential issue is not one of industrial conflict but is one of preventive medicine—a pragmatic field that deals with both the individual and the environment in any way that can decrease illness. From this perspective, screening for genetic differences in susceptibility to chemicals is a logical extension of the public-health practice of skin-testing for the presence or absence of acquired immunity to an infectious disease.

Second, it is artificial to question screening in terms of an abstract right of equal access to a job. One could hardly defend the right of a hemophiliac to be employed as a butcher. The problem is becoming more subtle today: We are learning to detect predispositions that are not so obvious to the bearer. Moreover, the problem is bound to grow. The growing fusion of genetics with immunology may soon make it possible to identify hereditary individual differences in the likelihood of developing various allergies. With such advances, certain illnesses cease being unpredictable acts of God. As they become increasingly predictable they also become more preventable—by minimizing exposure, and eventually, we hope, by learning how to modify an individual's reactions. Accordingly, we cannot escape the responsibility for incorporating knowledge of individual genetic differences in preventive medicine.

A third claim is that the limits of exposure to any chemical in industry ought to be set low enough so that nobody could be harmed. The moral principle assumed here is politically attractive. However, with our modern knowledge of genetics it cannot be taken literally. It can hardly be economically feasible to set standards on the basis of the responses of a person far outside the normal range of susceptibility.

Recent examples of mismanagement of toxic wastes hardly generate trust in the control of toxic exposure of workers in industry. Nevertheless, our suspicion should not blind us to the possibility that genetic screening could benefit both workers and management. There are real questions of policy in determining who should be responsible for the screening and how the results should be used, but an answer that would altogether reject the information would surely be unfair to the susceptible worker.

If we are to proceed rationally we must try to reconcile the ideal of unlimited equality with the reality of genetic diversity. We must therefore evaluate with particular care any arguments that may be used to discourage the useful recognition of that diversity. As the total suppression of genetics in the Soviet Union for twenty-five years by Trofim Lysenko reminds us, this field is particularly vulnerable to ideological assault.

Part Four

Medical Education and Affirmative Action

21

Trends in Basic Medical Sciences

This section starts with a talk, at a conference sponsored by the National Board of Medical Examiners, in which I discussed the decline in standards in medical education. A major factor was the demand from students, during the egalitarian revolution of the late 1960s, for more relevance in the curriculum, coupled with less emphasis on science and on the elitist goal of excellence. This paper is primarily addressed to the medical profession; but since it emphasizes standards, at a time prior to the conflict with affirmative action, it may shed light on the concerns that later motivated me to publish a controversial editorial, which follows.

I would like to make some general comments on medical education and then discuss our recent curricular experiment at Harvard Medical School.

It hardly needs stating that the unusual pressures of our times are having a large impact on the basic sciences, with much emphasis on innovation (one often feels for its own sake) rather than on excellence. Moreover, the society around us is trying to correct the gross inadequacies in the distribution of medical care by a crash program aimed at increasing the numbers of physicians and the relevance of their teaching—again without much concern for excellence.

This is really an extraordinary development, against the background of the flowering of the biomedical sciences in this country since World War II. I would like to quote from an editorial by Arthur Kornberg in the PHAROS of the ¢θ¢ for March 1971:

The National Board Examiner, Philadelphia, Pa., April 1971.

> Some scientists objected to this new scale of support of science almost from the outset. . . . This rising tide of support would populate science with mediocre people and inundate the literature with trivial data. On the contrary, the results of the massive support of science in the United States during the past twenty years have exceeded even the most optimistic predictions. Technology advanced far beyond our expectations. No one imagined that we would acquire so quickly the firm grasp we have today of the basic designs of cellular chemistry and its regulation. . . .
>
> Despite the spectacular success of this scientific effort, there is now an increasing retrenchment of support for research and training of scientists. I never expected this reversal of support. What I failed to anticipate too was that public apathy or hostility to science would be evident so quickly among scientists themselves.

The main point in this editorial is that we on the faculties have a real obligation to promote high intellectual and professional standards, and it is surprising how some are responding sympathetically to the attack on these standards.

Speculating about the reasons for the current broad attack on science in our society, I would first note the widespread frustration at encountering unexpected damaging side effects of technological advances, after having nursed the illusion that we would get pure benefits and no losses. For example, nothing could look like a purer advance for humanity than chemotherapy, ridding us of many infectious diseases. Yet this advance is largely responsible for the overpopulation that is now threatening our environment and contributing to so many of our problems.

Second, there is the loss of confidence in a society that has no general consensus with respect to its goals and values and can't even stop itself from committing increasing genocide in Vietnam today.

Third, the admirable concern of young people with social injustice is unfortunately generating pseudoegaliterian objections to excellence; for excellence implies difference.

Finally, the acute problems of medical education arise most of all from the failure of organized medicine and of society as a whole, particularly Congress, to reorganize medical care over the past decades.

Against this background we have developed a wave of irrational answers to very real and pressing problems. It's easier to blame the medical schools than to blame our infatuation with "free enterprise." This wave has even engulfed the recent Carnegie Commission report on medical education, which embodies all the current clichés and hardly reflects the intellectual distinction of the men who lent their names to it. The wealthiest nation in history, with 145 physicians per hundred thousand people, has less favorable health statistics than Sweden or England, with about 85 physicians per hundred thousand—and we are now told that a crash program to increase our ratio to more than 145 is required and will solve the problem.

So much for generalizations. Now let me talk about some concrete recent experiments at Harvard Medical School, which illustrate the contrast that so often develops between aims and consequences in large-scale experiments in education.

Two years ago our school shifted, like many others, to a core curriculum. Of the several arguments for this change the main one was that we needed to be more like a graduate school, and could do so by giving the students one and a half years instead of two of basic sciences, followed by a half year of electives. In the core they would be taught the necessary vocabulary and would get acquainted with the various subjects. Then each would choose the subject that excited him most and would learn the scientific method through a half year of intensive exposure to basic science in that area.

What has actually happened? First of all, for many teachers of basic sciences it has been demoralizing to be forced by outside pressures to make their teaching more superficial—on a large scale, in an untested experiment, and against their judgment.

I won't discuss student reactions because they seem so variable.

When our first class reached the end of that year and a half and were offered a tremendous variety of electives, about 75 percent of them asked to go right on into the clinical teaching of the third year. Thus within less than three years these students will have completed all the requirements for the degree, except for the number of hours put in. Our catalog is filled with elective courses that have no students. So with the aim of increasing electiveness in the curriculum, we have achieved for the bulk of the class a rigid curriculum that will probably soon be a shorter curriculum.

What is perhaps most relevant is that when this class took its National Board exams in the basic sciences, our record in most subjects dropped markedly from its level in previous years. Some would minimize the significance of this result by assuming that the core curriculum had discarded much irrelevant information that was still tested in the National Board examinations. But having taught and having been a member of the Test Committee for Microbiology both before and after the change, I know that at least for my subject this explanation is not correct. Though the hours allotted had been decreased by one-third, the material in the examination had all been covered; but the students had not learned it as well. The results provided by this external yardstick must be taken seriously.

I cannot pretend to be able to define accurately the many factors in this complicated situation. One is surely the distraction of much of the class by political activity and unrest. But I would suggest that the most important factor was the loss of incentive to take the subject matter seriously. Instead of absorbing the core curriculum avidly because it would so obviously be relevant and essential, the students reacted to an atmosphere that emphasized their good fortune in having to learn so little, rather than so much. Moreover, some of us found ourselves apologizing for occasionally introducing material that illustrated a scientific insight rather than tools of the trade. What an advance toward a graduate school atmosphere!

After two years I still found our teaching in the core curriculum an unsatisfactory compromise. It was too extensive for students who weren't at all interested in science, and who even feared that science might distract them from their goal of solving social problems. Nor was it satisfactory for students who wanted to learn all that they could in medical school. I then decided to offer a personal course as an alternative to the microbiology in the core. The students were told that they would have to do twice as much work in this course, that they would hopefully be able to learn twice as much, and that they would get no extra credit.

Twenty-three students out of 140 took the chance. I've never enjoyed teaching so much. I haven't felt it necessary to apologize for the irrelevance of anything that I was teaching. These students wanted to learn all that I could teach them and more. And what they apparently enjoyed most was the chance to consider evidence critically, and to reason, rather than simply to memorize the core material. This experiment thus revealed the existence of a silent minority of students who had been very quiet in our student body. Two other departments have now decided to offer more intensive alternative introductory courses for students who wish to learn all the science they can.

Where do we go from here? Even for those of us who do not consider the core curriculum satisfactory, it is an important experiment on which we should build. Surely we will not give up the effort to design a flexible curriculum to meet the needs of students with heterogeneous backgrounds and aims. In the long run, we may differentiate some groups who would receive a different degree in the health sciences; we could thus provide a more tailor-made education without worrying about the license to practice medicine in general. In particular, with the growth of the behavioral sciences, perhaps psychiatrists will eventually give up their emphasis on the need for a full medical training to provide legitimacy and insure high professional standards. If they become willing to encourage a different educational process and professional degree for psychotherapists, I believe much of the tension over the role of the basic sciences will be eased—for most of the material in these sciences is indeed much less relevant to psychiatry than to any of the other branches of medicine, and much of the protest comes, legitimately, from future psychiatrists.

Meanwhile, my hope is that we will move away from seeking an ideal standard core and will develop a truly elective system—not a delayed elective after a year and a half or two, but an elective from the start, with a multi-track system. This system is the heart of the kind of experimentation we are going to have to undertake if we don't want to shortchange those students who want to take science seriously. And we have been shortchanging them.

I don't know what we're groping toward in medical education. We may soon be giving mostly three-year degrees; or we may be giving four-year degrees with a much more elective curriculum, so that the future psychiatrists will spend more time in behavioral sciences and even in literature, rather than in bacterial genetics or gross anatomy. And we will clearly also have more M.D.—Ph.D. students. But my feeling is that it would be a shame if the only choice for the student should be a scientifically skimpy three-year degree or a six- or seven-year M.D.—Ph.D. degree. We should also aim at multi-track programs in which a student in a four-year course can also have at least as good a scientific training as most of us have been exposed to in the past. If we do not, and if we convert our medical schools into trade schools, they will surely attract fewer outstanding people, both as teachers and as students.

Finally, while recognizing a serious problem of medical care in our country, I would like to express my conviction that the solution does not require, and would not even benefit by, weakening the fantastically successful biomedical research enterprise that is unique to our medical schools. It is important to preserve this enterprise, and the sophisticated teaching that it promotes—not only to enable potential future medical scientists to become interested, but also because man does not live by bread alone. However pressing our problems of distributing bread and distributing medical care may be, we also have a culture to preserve. Scientific research is among the most creative cultural activities in our world today, comparable to cathedrals in the Middle Ages and painting in the Renaissance. It would be a shame if our effort to solve pressing current social problems should lead us to pull the whole thing up by the roots and then hope that some day it will start growing again.

22

Academic Standards in Medical Schools

The following guest editorial in the New England Journal of Medicine *was picked up by news media throughout the country and quite a storm developed. An essay recounting the history of that turbulent episode follows thereafter.*

Since the consumer is particularly blind in purchasing medical care, and his vital interests are often at stake, those who are in a position to screen for aptitude and competence in medicine have a grave moral responsibility. In accepting this responsibility medical faculties have always taken into account qualities of character and motivation as well as scientific ability and knowledge. In addition, in recent years we have finally begun also to take into account long ignored social needs. But no one of these sets of qualifications can compensate for a gross deficiency in another. In particular, as the practice of medicine broadens its scientific base it increasingly requires a reasonable level of competence in science, at least as long as the M.D. degree leads to an unlimited license

---

New England Journal of Medicine 294 (1976):1118.

to make life-and-death judgments. In this connection preclinical courses serve not only to provide a scientific background for practice but also to screen students for the ability to reason scientifically.

This screening has become more difficult in recent years. A variety of considerations have led medical schools to engage in innovations in admissions, curriculum, grading and criteria for promotion. Some faculties, no longer confident of their ability to maintain adequate minimal standards, have set an external standard by requiring candidates for the diploma to have passed Parts I and II of the National Board Examinations. But for schools that have aimed at leadership this minimal national standard is an extraordinarily low one. Moreover, it has been further lowered in recent years: National Board grades are normalized for each year's population, and so the absolute norm for passing is necessarily lowered by any nationwide increase in admission of students with substandard academic qualifications.

It would be a rare person today who would question the value of stretching the criteria for admission, and of trying to make up for earlier educational disadvantages, to help disadvantaged groups. But how far faculties should also stretch the criteria for passing students is another matter. If a board licensing airline pilots allowed extraneous considerations to interfere with objectivity it would be considered criminal. The temptation to award medical diplomas on a charitable basis raises the same question, even though the consequences of fatal error in the two professions are not equally visible and dramatic.

Many faculty members have wondered whether the stretching of standards in their schools in recent years has not exceeded what is reasonable. The problem is illustrated by a distinguished school that recently waived its National Board requirement and awarded a diploma to a student who had been unable to pass Part I in five tries. The award of this degree was virtually inevitable, after five years of investment by the school and the student. But we must look at the erosion of internal standards, and the postponement of decision, that allowed this situation to develop.

Medical faculties can derive deep satisfaction from their success in recruiting and helping many able students from groups that were formerly excluded. But it has also become apparent that patience and sympathy cannot overcome the inability of some students to handle the material. It is cruel to admit students who have a very low probability of measuring up to reasonable standards. It is even more cruel to abandon those standards and allow the trusting patients to pay for our irresponsibility.

Considerations of tact, and guilt over our history of enormous racial injustice, have made it difficult to face the problem. But there are dangers in a policy that fails to evaluate the results of our recent experiments objectively. If the public is given a romanticized view we can expect demands for the extension of quotas, rather than demands for strengthening the quality of the product.

It seems time for medical faculties to ask whether we have been properly balancing our obligation to promote social justice with our primary obligation to protect the public interest, in an area in which the public cannot protect itself.

23

Affirmative Action and Veritas at Harvard Medical School

I did not write up this history earlier because I thought it would have been too embarrassing to my university and to many individuals. But it seems appropriate to present the story now, in a calmer atmosphere. Future historians deserve documentation on the cover-up of facts, and the intimidation, that prevailed during the mid-1970s in our universities over their affirmative action programs. Moreover, our society is still wrestling with the underlying problems. I hope the other participants in the controversy at Harvard Medical School will find these recollections accurate and will accept my reasons for presenting them.

Unlike the other pieces in this volume, this one has not been previously published.

The Background

Like most academic institutions, Harvard Medical School, while priding itself on independent, free inquiry, has inevitably adapted to the prevailing social prejudices and fashions. In the early decades of this century it had a de facto quota for Jews; it did not admit women until the 1940s; and only rarely did Negroes enter. In 1968, after the murder of Martin Luther King, Jr., the faculty took the dramatic step of voting to admit a minimum of fifteen blacks, even though the pool of qualified applicants was known to be small. Some members argued passionately that if we tried hard enough we surely should be able to find fifteen satisfactory candidates. The argument seemed reasonable, but it ignored the fact that virtually every other medical school in the country was trying to tap the same pool.

This remarkable faculty action reflected the drastic transformation that our political system had imposed on the concept of affirmative action. Originally specified by Congress as a program to eliminate discrimination and to remedy previous educational deprivation, affirmative action had become a program of enforced quotas. The underlying assumption was that the terrible legacy of slavery obligated us to short-circuit for blacks the gradual path that had been followed by other groups who started at a disadvantage: advance in education, then achievement, then social status and income. Instead, it was argued that racial justice must be defined in terms of rapidly reaching proportionate numbers in all occupations, rather than in terms of equal opportunity to develop and compete.

Harvard Medical School accepted this prevailing dogma. It was argued that among minority students who lacked proper academic qualifications many would have potential that had been buried by previous discrimination, and so they should be able to do well once they were afforded the opportunity. But this was a gamble, and so at the faculty meeting that voted in the quota I asked whether we would also plan to lower our standards for passing courses. The dean, Dr. Robert H. Ebert, replied that there had been no mention of lowering standards, and he had no intention of letting that happen.

The performance of the early students was in fact disappointing. The administration then decided that our Admissions Committee was not well suited to recognize buried potential in applicants from an unfamiliar cultural background. Accordingly, a special Minorities Admissions Subcommittee was set up, with a predominantly black membership (including many students), and it was soon allotted approximately 20 percent of the entering class. Dean Ebert's thesis was that Harvard was in this way continuing its tradition of leadership. In an earlier age the main goal was to set standards of academic excellence and to train leaders in American medicine; but now the school must also set an example of leadership in graduating minority physicians. But this approach, however well intentioned, ignored an obvious dilemma: leadership in academic excellence inevitably made it more difficult to absorb students who lacked, for whatever reasons, a good academic background.

Unfortunately, even with the new subcommittee a substantial fraction of the minority students continued to fail to meet our earlier standards. We could thus no longer build on the assumption that vigorous recruitment would solve the problem. Faced with the choice of either abandoning quotas or drastically lowering standards, our school chose the latter—though not by open faculty debate. The dean's office simply pressed the departments to provide repeated reexaminations to failing students, and inevitably these examination became less demanding. Virtually every student eventually passed each course. (Incidentally, this policy had an unintended by-product: since the school could not justify a double standard for passing, its lowered standards provided a cushion for some nonminority students whose performance would not have met the medical school's earlier standards.)

Obviously, for those students from poor educational backgrounds who showed continued improvement a good deal of patience was clearly justified. But the patience was extended much farther: given the pressure to enroll and graduate large numbers of minority students, and the inevitable reluctance to discard a cumulative and expensive educational investment, the hope for eventual improvement often became an excuse for not dropping a student of limited capacity. And to justify this practice, the quaint notion was advanced that even if some students took longer than others to pass a course, or if they required five years or more to complete the four-year program, they could then be certified as being just as qualified as those students who had mastered the material more readily.*

The Justifications for Lowering Standards

Several arguments were used to rationalize the new attitude toward standards. One was that skill in taking examinations was culturally conditioned, and thus our examinations were not fairly measuring the minority students' ability or knowledge. Another argument arose from the fact that these students had their greatest problems in the early, basic science courses. It was therefore suggested that the long tradition of building on these courses as a foundation for clinical training might have been wrong: perhaps one really did not need to be competent in science in order to be a good physician.

Indeed, the amount of basic science that is appropriate in medical education is a perennial and legitimate question. It is clear that effectiveness in some areas of medicine, such as psychiatry or plastic surgery, depends primarily on talents and knowledge that are far removed from biochemistry and microbiology. But as long as the graduate is free to choose any branch of the profession, it is difficult to see why a minority program justifies abandoning the otherwise acceptable requirement of a comprehensive background in science. Moreover, the basic science courses have always played a useful screening role in providing an objective evaluation of a student's ability to learn and to reason, since the later evaluation at the bedside depends much more on personal interactions and both sets of qualities are important. While one need not show gifts as a biochemist in order to become an excellent psychiatrist, I would worry about the candidate who simply did not have the intellectual capacity to handle this material at a minimal level.

Perhaps the most compelling argument was that we must help, even at the price of moderately lower standards, to meet a desperate need for

*While the example of Dean Ebert's dedication to keeping up the numbers no doubt influenced many other medical schools, this policy was not required by the law, or by any overwhelming social forces, and it was not universally followed. Johns Hopkins Medical School, for example, having also set up a program in the late 1960s based on quotas, soon dropped it rather than lower academic standards. They could later build up the number of minority students without sacrifice of quality by assuring minority candidates with academic promise that the value of their diploma would not be diluted. Alternatively, some schools offered minority candidates a year of special education between college and medical school to help remedy educational deficiencies and to provide an opportunity for testing (and self-testing) before commitment to a medical career. This policy seems kinder than Harvard's "cold turkey" entry, and more realistic.

physicians in minority communities. This argument tacitly assumed that a student coming from the ghetto was likely to return there, even though a Harvard diploma would open up many other avenues. Indeed, on this assumption, bolstered by political convictions, the Minorities Admissions Subcommittee sometimes rejected well-qualified middle-class applicants, while accepting less qualified ones from what they considered the proper background. (More on this later.)

The administration took several steps to make the radical changes in academic standards less conspicuous. First, letter grades were replaced by a system of either pass or incomplete, rather than pass/fail; and when a student replaced an "Incomplete" by "Pass" his record retained no evidence that he had had difficulties. This change in grading came in as part of a more general experiment in curricular reform, but it provided the dean with a convenient device: he could honestly state that the performance records of the minority graduates could not be distinguished from those of the other graduates.

Another move more deliberately deprived the faculty of objective feedback on student performance. In the past the ranking of our students in the National Board Examinations, in each subject, was presented each year at a faculty meeting, and any department that fell below third place in the country virtually apologized. Shortly after the new admissions program started the dean's office quietly dropped this annual report; and I have only unofficial information that our national standing has become much lower.

To be sure, it is not clear how much the changes in admissions and in grading contributed to this decline, because we had meanwhile changed to a largely elective curriculum. The important point, however, is that the faculty has received only one post-mortem on a large and prolonged educational experiment. This report, comparing the performance of the minority and the other students in the National Board Examinations in two selected classes, showed improvement, but still a distressing number of failures, in the minority group. But this revelation of a serious problem did not lead to periodic follow-ups. In a sense, then, the faculty still functions somewhat like those medieval surgeons who are alleged to have used a concealing sheet, for reasons of modesty, when operating on female patients.

Because the faculty thus found it very difficult to take responsibility for failing a student, it voted to establish some kind of cutoff by ruling that the requirement for our M.D. degree would include passing the National Board Examinations. While this cutoff somewhat relieved our consciences it did not offer very stringent protection, since it settled for the minimal national standard. Moreover, a student could repeat the examinations up to five times. But even the modest National Board requirement was subsequently waived, under the following circumstances.

Private Memorandum, Public Editorial

Early in 1976 a colleague came to me to suggest that I might be interested in questioning a technically illegal action that the dean was planning to slip through the next faculty meeting, involving a black student who had failed the basic science part of the National Boards for the fifth time. (He had passed all our courses, requiring an extra year.) The university statutes require that all diplomas be awarded by a vote of the faculty; but the dean had arranged for our Administrative Board to waive the National Board requirement and to vote the award of a diploma.

Announcing that waiver at the faculty meeting, the dean stated simply that the student had failed the National Board basic science examinations but had satisfied all our other requirements. Moreover, the student, who had come to us from West Point, was now a medical officer in the Army, but he would soon revert to being a line officer unless he received a medical diploma. The dean, who had served as faculty adviser to this individual, assured us that he was a fine student with an excellent record in his clinical work. However, the dean was not involved in teaching, and I later found that his evaluation was sharply contradicted by colleagues who had taught this student.

My informant had picked the right proxy. At the faculty meeting I asked whether it was not an undesirable precedent to have the Administrative Board replace the legally required faculty vote on a diploma. To resolve an embarrassing situation someone moved that the faculty vote the diploma. In the brief discussion that followed another member innocently asked why the student did not take the examination again, and the dean stated simply that he had done so. The faculty then voted favorably, without having been informed of the five failures. (In his subsequent public defense of the school's standards the dean referred to an "overwhelming" vote of the faculty.) At that faculty meeting, and in later public controversy with the dean, I did not have the heart to reveal this bit of academic trickery—but now, in presenting an unvarnished story, I cannot avoid it.

I was troubled to see how far the virtuous aim of trying to meet affirmative action goals was in effect distorting the tradition of *veritas* in the university. But since this student had passed all our courses, I was even more concerned by what this episode implied about how far we had relaxed our own standards. I therefore drafted a memorandum to the Faculty Council, which was cosigned by six other senior faculty members, arguing that this case illustrated the urgent need for a better balance between our effort to redress past social injustices and our obligation to graduate only competent physicians.

The memorandum led to a regulation limiting the number of makeup examinations the medical school would allow. A number of colleagues, and the dean, commented favorably on my formulation of a troubling problem. Since many other medical schools were wrestling with the same problem, I decided to publish a very similar statement, without identification of our school, as a guest editorial in the *New England Journal of Medicine* (see previous essay, pages 168 to 170). And to avoid any involvement of those colleagues who had cosigned my memorandum I did not include their names.

I might note that a year earlier another guest editorial in the same journal, by a professor of pathology at the University of Kansas, had directly criticized admissions policies for minorities, and that piece had evoked only a few angry letters in the journal. Since my article focused primarily on standards for passing all students, and only secondarily on the influence of revised admissions policies on those standards, I did not expect much reaction.

The Response in the Media

My first contact with the news media was not reassuring. A medical writer for the *New York Times*, Dr. Lawrence Altman, wrote a balanced article, which recognized that a growing number of medical teachers were privately expressing criticisms like mine. Because the issue was so delicate he took the unusual step of reading the article to me before publication, and I had no objection. But while his submitted draft cited a favorable statement in my editorial (one of five such) about the goals of affirmative action, an editor removed that statement; hence the published story presented a simpler and more sensational picture, in which I appeared to be expressing only criticism of the program, and no sympathy.

Subsequent news articles elsewhere further simplified the story, and they did not mention that any other medical teachers shared my concern. Many quoted a statement, in my interview with the *Times*, that graduation of an incompetent physician was likely to result in "a swath of unnecessary deaths." This statement aroused a strong reaction. I still believe it was a simple and obvious truth, but I would now say that in the context of the tensions over affirmative action it was no doubt too dramatic.

A really inflammatory story appeared in the local student newspaper, the *Harvard Crimson*. I did not try to correct distortions in other newspapers, but because the *Harvard Crimson* is widely read in the university community in which I live I sent it the following letter (published May 19, 1976).

To the Editors of *The Crimson:*
I must protest the slanted nature of the article by Judith Kogan (May 14) on my recent piece in the *New England Journal of Medicine.* The issue is a complex and delicate one, and when I learned that *The Crimson* was preparing to run a story on the matter without even having seen my article I personally delivered a copy that evening, suggesting that reprinting it would convey my message more accurately than a set of paraphrases and selected quotations. Instead *The Crimson* started with an outrageously inflammatory headline (Professor Assails Blacks' Performance) and then quoted exclusively the critical aspects of the article. It ignored the parts that made clear my support for minority programs and my desire to see them strengthened by resisting pressures to stretch standards excessively. Thus ". . . medical faculties have always taken into account qualities of character and motivation as well as scientific ability and knowledge. In addition, we have begun also to take into account long ignored social needs. . . . It would be a rare person today who would question the value of stretching the criteria for admission, and of trying to make up for earlier educational disadvantages, to help disadvantaged groups. . . . Medical faculties can derive deep satisfaction from their success in recruiting and helping many able students from groups that were formerly excluded. . . . Considerations of tact and guilt over our history of enormous racial injustice have made it difficult to face the problem. But there are dangers in a policy that fails to evaluate the results of our recent experiments objectively."

The Crimson has thus created the false impression that I am criticizing the performance of black students as a whole, instead of emphasizing the need to distinguish a satisfactory from an unsatisfactory student, regardless of ethnic origin. By so distorting the picture *The Crimson* has injured the black community, and also those (including me) who are sympathetic with their needs and aspirations. Indeed, I do not blame anyone for getting angry at my views as portrayed by *The Crimson.*

It is important to correct not only that picture but also any possible connection between the content of *The Crimson* article and the views of Professors Amos, Anderson, Hubel, Karnovsky, and Rosen. They cosigned the original document, prepared for the Faculty Council, that was the basis for my published article, and no statement by me outside that document should be ascribed to them. I apologize for my indiscretion in identifying those colleagues: I felt free to do so since the document is scheduled to be distributed at a faculty meeting, which students can attend. . . .

The original document was accepted without criticism by the Faculty Council, which unanimously passed two resolutions addressed to the problems. It is thus clear that these problems are widely perceived, by educators close to them, as real and significant. Because this formulation had proved so useful I submitted to the *New England Journal* a condensed and updated version, intended as a reflective comment for consideration by medical educators at other schools. In this article I did not criticize my school—indeed, I am very pleased by the progress we are making. In particular, I did not identify Harvard as the school that had finally awarded a diploma to a student who had failed Part I of the National Boards five

> times. (I would now like to add that the recommendation for a late award of the degree was based on evidence of subsequent satisfactory clinical performance.) I specifically asked the reporters from the *New York Times* and from *The Crimson* not to identify Harvard Medical School in this connection. A reader of my article could, of course, make a reasonable guess—but I hope there is still a place for tact in discussing such issues. The *Times* honored my request; *The Crimson* did not. I apologize to the administration of Harvard Medical School for the result of my indiscretion.
>
> I am very sorry that statements quoted in the press may have led minority students to believe that I have been criticizing their performance as a group. I trust the original document will make clear my recognition of the fundamental success of minority programs in medical schools, and my concern for ensuring good medical care for all segments of society.
>
> Now to the most serious matter of all. . . . Lewontin's comments. He is quoted as saying "[Davis] thinks blacks are mentally inferior and incompetent. . . . [He] argues that these minority students don't have the intrinsic ability to become doctors." Nothing in my article justifies this grave charge. Neither does anything else that I have said or published. I have written to Professor Lewontin demanding an immediate and full retraction.

Incidentally, Professor Lewontin refused to retract the scurrilous statement noted at the end of this letter, and my legal counsel advised that in the political climate surrounding this issue a suit would be very hazardous. Though Lewontin is a sophisticated population geneticist, he deliberately ascribed to me a racist, typological position, lumping together all the members of a group as a justification for treating them differently. Yet he knew very well my quite opposite views: that genes as well as environment contribute substantially to the observed differences in abilities between individuals; very likely both also contribute to differences in the distribution of various abilities in different races; but since the latter differences are statistical rather than typological, with overlaps between all groups, they should not influence our treatment of individuals.

While the article in the *Harvard Crimson* had a local impact, a subsequent article in the *Boston Globe* was the probable basis for the extensive coverage by the national news services. The *Globe* reporter, Richard Knox, had failed to write up my editorial on the day when it came out, so he had been scooped by the *Times*—and on a Boston story. The next day he interviewed me by phone for a full hour, and I naively answered his probing questions as candidly as I was accustomed to doing with students. The resulting long article gave the false impression, through careful selection of material, that I was opposed to any effort to help black students. Moreover, he quoted statements from many members of the Harvard Medical administration and faculty virtuously condemning my action. Only one colleague, Dr. Sargent Cheever, was supportive, saying that he believed I was truly concerned about standards and was not acting on racist beliefs.

On the morning when this article was published two television stations sent teams to interview me. Both reporters were black. The first was extremely hostile, asking me repeatedly whether I believed that black folk were inferior; he was not interested in my editorial. The second reporter handled the interview in an entirely fair manner, which I found quite admirable under the circumstances. He read the editorial while the cameras were being set up, then whistled and said that this was very different from the story going out over the national news services. Only then did I realize how deep a hole I was in.

After the story became national news I received hundreds of private expressions of support from colleagues, at the school and elsewhere. In the fever of public denunciation, however, it would have taken a great deal of courage to offer any public support. I felt that the editorial provided my best defense, and I was pleased that the *Boston Globe* and the *Harvard Crimson* reprinted it; the OpEd editor of the *New York Times* refused to do so.

Some students held a rally, and they picketed me briefly. The Associate Dean for Minority Students, Dr. Alvin Poussaint, made the main speech. I had hoped that he would play a dean's role and try to calm the students; and since he was familiar with my record I had asked him to try to correct their misperception of my motives. However, his speech proved to be that of a politician appealing to the emotions of his constituency. When asked on television that evening whether he thought I was a racist, he said, "Well [long pause] he says he is not a racist."

A professor of psychiatry who was also an associate dean had suggested that it might be unsafe for me to be in my office at the time of the rally, and he invited me to watch it with him from the dean's office, where I could safely meet with reporters afterwards. During the meeting he suggested that the atmosphere was getting pretty heated up, and it would be wise for me to go home. I meekly accepted his collegial advice. I later learned that the reporters had been disappointed at not finding me in my office after the rally, and only then did I realize that I had been maneuvered away from an opportunity to give them my rebuttal to the charges that they had just heard.

The Dean's Responses

Faced with a flood of inflammatory stories in the media, Dean Ebert released the following statement to the general press, as well as similar statements in the *Harvard Crimson* and the *New England Journal of Medicine*.

> Both the faculty and administration are certain that all of the students granted the M.D. degree are highly competent and will make excellent physicians.
>
> I know of no evidence to support the view that the students at the Harvard Medical School have diminished in quality in recent years. Indeed, I would say the standards are as high as they have ever been—perhaps higher.
>
> Dr. Davis, in publishing his article and speaking to the press, speaks only for himself and not for the administration of the Harvard Medical School or the rest of the faculty.
>
> I believe that Dr. Davis's action in identifying an individual is irresponsible, since there was no way of answering the charges without revealing more information on a matter which had been handled internally by the appropriate committees.
>
> In actual fact, the case was a unique one. The student was awarded his M.D. degree only after exceptional proof of his clinical competence. The faculty then voted overwhelmingly to grant him a degree.
>
> Dr. Davis's statement was also irresponsible because of the general implications about the professional acumen of all minority students.

I have already described the circumstances of the "overwhelming vote." I would further note that in obtaining that vote the dean had not claimed "exceptional proof of his [the student's] clinical competence," let alone provided evidence.

The dean also sent a memorandum to all heads of departments stating that "all information with respect to students and other faculty matters is confidential and is not for public release without specific permission from the Dean of the Faculty of Medicine." Whether I had betrayed a trust, as implied by this statement, or had engaged in justifiable whistle-blowing, I would note that the confidentiality of faculty meetings did not weigh heavily on members, for ever since the campus unrest of the 1960s students were allowed to attend these meetings.

In addition to his barrage of public statements, the dean obtained statements supporting his position from the Faculty Council and from the committee of chairpersons of the preclinical departments. I can understand the need for a vigorous response, but I believe that even in the heated atmosphere then prevailing, a courageous dean, dealing with students who were understandably outraged at what they read in the papers, could have sympathized and at the same time tried to convince them to look at the editorial itself, and at my record in the school. Unfortunately the course that Dean Ebert chose erased any possibility of correcting the public impression that my criticisms of his program were a racist attack on affirmative action.

The department chairmen who publicly supported the dean were all good friends of mine, and before their statement came out one of them phoned me to express regret for this action, which he considered necessary to restore good relations with the students. While it did not surprise me that colleagues would run for cover when a public storm arose over views that some of them shared, their formal public censure was the most disappointing aspect of the whole episode. I have no way of knowing whether this behavior was based primarily on concern for the anguish of the minority students, on the desire to avoid bad publicity for the school, or on fear of being tainted as racists.

President Derek Bok, of course, was obliged to comment on the questions that the press raised. He discreetly stated that "On the basis of the evidence supplied to me by the dean's office and the registrar, I find no basis for any implication that minority students are less than fully qualified for the M.D. degree in accordance with the normal standards of the Harvard Medical School."

After making the public responses that I have noted, Dean Ebert took a remarkable further step. He sent a letter to the dean of every medical school in the country, denying that there had been any lowering of academic standards at Harvard, and expressing the hope that their admissions committees would not be influenced by my irresponsible actions as evidenced by the editorial. In addition, he enclosed the supporting statements that he had obtained from the Faculty Council and the preclinical department chairmen, thus fostering the impression that my position was not shared by any colleagues.

Finally, the dean informed me that under the circumstances he could not proceed with his earlier plan to make me director of the Center for Human Genetics. He apparently had forgotten that he had given me that appointment a few months earlier. I had accepted this minor post (which entailed presiding over the division of a fellowship grant among several groups in the school) purely as an administrative service, and I derived no real benefit from it. But the principle of being dismissed on political grounds was not trivial. Reporters had earlier asked whether the school had altered my official position in any way, and I was glad to be able to say no. I therefore visited President Bok to offer to resign from this particular assignment, or to accept whatever other solution he considered least embarrassing to the school. The matter ended with my being invited to choose the solution, and I elected to retain the position but with a change in title.

The dean later told me that he had undertaken to cancel this appointment at the urging of a group of faculty and students, headed by a geneticist whom I had known, on a friendly basis, for thirty years; and, of course, this person was to take on the post. The dean also remarked that another group of students had visited him to demand my dismissal, and he had vigorously defended the principle of academic tenure. In recent years this principle has been widely questioned, since it is now rarely called on to serve its original purpose of preserving the right to free inquiry and expression; hence it is worth noting that my experience provides a rare test of its continuing value.

The Apology

Early in this hectic period a member of the dean's office told me that the news stories about my article had led some patients in our teaching hospitals to refuse to be treated by black medical students. I was horrified, and I suggested to the dean that the most effective way to correct this misinterpretation might be for us to avoid making opposing statements and to issue instead a joint statement in which we agreed that my criticisms applied only to a few individuals and not to the black students as a group. He said that I could do whatever I wished, while he would do what he felt he had to.

I therefore released the following statement, which was widely noted in the news media.

Statement by Bernard D. Davis on His Article on Academic Standards (Press release, Harvard University, May 21, 1976)

> I deeply regret my failure to anticipate that my article in the *New England Journal of Medicine*, intended for professionals, would reach the public press. Its misinterpretation by some of those who have commented on it publicly, together with ill-considered subsequent responses of mine to queries from certain newspapers, have caused much harm. For my share of these errors I apologize. My article did not raise any doubts about the quality of minority students or physicians as a group. I do not have such doubts, and it would be utterly contrary to my convictions, both personal and scientific, to make any such generalization about any racial or ethnic group. My only concern is with a very small fraction of the students, both nonminority and minority.

I might add two comments on this statement. Dean Ebert told the press that he had reacted so forcefully because my charges had led so many patients to refuse black students. However, several clinical department chairmen with whom I checked did not know of any such incidents; and later, when matters calmed down, a knowledgeable member of the dean's office privately admitted that he could not substantiate the claim. Nevertheless, I do not regret having published the apology, since the picture appearing in the news must have stimulated such concern among patients, and it clearly caused a feeling of great hurt among the black students. Yet a basic question remains: if our school (and others) had indeed been passing truly unqualified students, and if the political climate made it impossible to rectify this policy without bringing it out in the open, was there any way of doing so without hurting innocent black students?

The other comment is that my statement was carefully worded as an apology for statements to the press, and not a retraction. But many readers no doubt missed that distinction. Thus a stranger sent me an interesting telegram, consisting of the words allegedly whispered by Galileo: "Eppur se muove."

Outside Attacks

Not surprisingly, other parts of the medical establishment supported Dean Ebert's actions. In an unctuous editorial the *Journal of the American Medical Association,* well known as a defender of orthodoxy, joined in denouncing my criticisms. Similarly, in a letter to the *New York Times,* the president of the Association of American Medical Colleges suggested that the excellent average performance of recent medical classes disproved my claim about lowered standards. Since the actions that I criticized involved only a small fraction of the student body, and hence would have little effect on the national averages, this argument seemed to me to be clouding the issue. I therefore published a reply, part of which follows:

Re: "Troubles in Medical Academe": A Clarification (letter to the Journal of the American Medical Association, *January 3, 1977):*

> *To the Editor:* In an editorial on July 26 (236:388, 1976), H. H. Hussey described my article in the *New England Journal of Medicine* (294:1118-1119, 1976) as a spring that ended in a dive rather than a vault. In support of this conclusion, he noted that (1) Harvard's dean had called my action irresponsible, (2) I had apologized, and (3) in the *New York Times,* President Cooper of the Association of American Medical Colleges had convincingly explained why my charges were not true.
>
> On the contrary, now that the initial furor has abated, it seems clear that my message has brought into the open concerns that are widely felt, at least among those who have teaching contact with students and not simply administrative contact with symbols. It is important to recognize these concerns about standards for graduation, for we must ensure that our minority programs succeed, and the greatest threat to their success, as well as to morale in medical teaching, has been our failure to maintain adequate standards in recent years. One could defend such a policy at the start of these programs as a means of priming the pump, but its continuation after eight years must give us pause.
>
> It would be unfortunate if the *Journal's* editorial defense of the status quo should discourage a constructive response to the airing of this question. I would therefore like to comment on Dr. Hussey's evidence.

1. It is true that Dean Ebert called my action irresponsible. That is a matter of judgment: it is understandable that he would see the immediate reaction as a threat to programs of minority education, though it remains to be seen whether my criticism of a vulnerable feature may not lead in the long run to strengthening those programs.

2. My public apology was occasioned by my learning that some patients who rejected black medical students were now citing newspaper articles based on my editorial. I was deeply disturbed by this reaction, and I felt obligated to try to neutralize it. But I did not, and do not, retract any part of my editorial.

3. Though Dr. Hussey found Dr. Cooper's letter in the *New York Times* (May 23, 1976) convincing, his editorial would have presented a more balanced picture if it had noted the following reply, which I published in the same newspaper on June 11:

> President Cooper of the Association of American Medical Colleges has rejected the charge that medical school standards have dropped in recent years (letter May 28). Indeed, he cites evidence for improvement in average qualifications and performance. But my editorial in the *New England Journal of Medicine,* which made the charge, was not referring to the average level—either of medical classes as a whole or of minority students.
>
> Rather, I was addressing the problem of minimal standards for failing an unsatisfactory student. The numbers involved are too small to have lowered the average quality of medical education and practice, especially with the extraordinary recent increase in the number of brilliant applicants for medical school. Nevertheless, even a few inadequate physicians are important, both for the patients whom they treat and for the image of the profession.
>
> It is clear that most minority medical students have performed very well. Indeed, they have earned admiration for their perseverance in overcoming early disadvantages. But the well-earned credentials of these good students may be tarnished, and the communities served will suffer, if poorly qualified members of the group are also passed. That is the thrust of my editorial. Neither that article nor any other statement of mine justifies mistrust of minority students or minority physicians in general.

Retractions and Corrections

A few months after the storm a thoughtful article by a graduating Harvard law student, J. W. Foster ("Race and Truth at Harvard," *New Republic,* July 17, 1976), accused the university of systematically lying (his word) in covering up the problems of its affirmative action programs. A similar article by W. Havender appeared later in *The American Spectator* (March 1978). And a few months after the outburst Michael J. Halberstam, a courageous physician and medical columnist (who was later killed while pursuing a burglar) wrote in *American Medical News* (December 13, 1976):

> By overstretching their standards, by pretending that a problem does not exist, by howling that anyone who questions minority recruitment is a racist, the defenders of such programs have put into doubt the qualifications of the vast majority of black and other recruited students who are clearly capable of excellent work. . . . Putting this responsibility into the hands of men and women who can barely read enough to squeak by graduation will haunt us 20 and 30 years into the future.

In contrast to these responses, the July/August 1976, issue of our alumni magazine, *The Harvard Medical Bulletin*, presented a history of the events (including the major published documents) that reinforced the dean's campaign to make my position appear totally isolated. Yet over a hundred colleagues on the faculty had meanwhile sent me private messages of support. I therefore told the editor, Dr. George Richardson (a friend and former student), that he surely must be aware of this strong division of opinion; and while I was willing to let calumnies in newspapers go by, I insisted that because his magazine provided a permanent historical record it must correct the picture. He replied that he could present only factual material, and not anecdotes about faculty opinion.

By then I had outgrown the naive idea that I was engaged in an intellectual discussion with colleagues, and I had become a bit tougher about the political battle in which I found myself embroiled. Accordingly, I offered the editor three choices: he or the dean could publish a statement correcting the distorted picture; he could invite me to publish my version in his magazine (which I preferred not to do); or he could do nothing and I would then publish the whole story elsewhere (the solution that I least favored, since at that time, with Dean Ebert still in office, it would have been very embarrassing to the school).

Meanwhile, some alumni were sending in criticisms of the dean's actions. Whether the editor was reacting to these or to my ultimatum, he was responsive *(Harvard Medical Bulletin,* November/December 1976). His statement took considerable pains to try to restore my personal reputation, and it included the following:

> . . . Prior to the outburst in the news media, then, Dr. Davis was clearly a professor operating together with his peers in an unequivocally worthy cause, that of academic excellence, particularly in the preclinical sciences that are his personal and professional concern. Davis's own credentials, furthermore, indicate a consistent concern with social justice: he was an organizer of antiwar demonstrations in the 1960s, the first department chairman in the history of Harvard to preside over the appointment of a black man to a tenured post, and for many years a member of the advisory board of the Civil Liberties Union of Massachusetts. As a teacher of genetics to undergraduates (Nat. Sci. 37), Dr. Davis is seen by the *Harvard Crimson* as providing a "strictly objective discussion of biology and genetics, laying the foundation for an analysis of the implications of recent advances in these fields for philosophy and ethics."

Dr. Richardson went on to say "We can only agonize as the two sacred cows of academic excellence and social justice gore each other in the public arena, with a third sacred cow, that of Dr. Davis's academic freedom, left bleeding on the sidelines."

Dean Ebert also made a corrective statement in the *Bulletin*, more restrained than the editor's, but nonetheless helpful. After admitting that he might have overreacted or have seemed unfair in his criticisms of me, he wrote:

> I know perfectly well that Dr. Davis is not a racist and I know that his commitment to academic standards is sincere and not an excuse to attack any ethnic group. I also know that Dr. Davis did not intend to undermine a policy of recruiting minorities for admission to HMS. Unfortunately, what we say and what we write can be misinterpreted, and my intent has always been to prevent what Dr. Davis has said from providing ammunition for those who wish to abandon our commitment to minorities. It has not been my wish to injure Dr. Davis or to prevent him from stating his views publicly.

I thanked the dean for this step to clear my reputation. But I was sorry to see his continued insistence, in other parts of his statement, that criticisms of our minority program must be vigorously opposed because they are bound to be used as an excuse for repudiating the recruitment program altogether. For opposition in his sense—which discourages debate—is of course antithetical to the ideals of academia and of a democratic society. Nevertheless, on the particular issue of affirmative action this view has received remarkably widespread support.

The same issue of the *Bulletin* also contained several letters on the controversy from alumni, mostly focusing on the question of academic freedom. One of these said:

> Harvard's actions, meant to strengthen public confidence in minority M.D.'s, have instead invited the inference that the minority program at Harvard survives by intimidation of potential critics. On any other aspect of academic affairs, a faculty member might be thought to have not merely the right but the duty to express himself as Dr. Davis has. For doing so, however, Dr. Davis has been flogged through the fleet, so to speak, in a dean's letter to 118 medical schools, with the preclinical chairmen and Faculty Council in attendance as witnesses. Such treatment of a faculty member, amounting to official censure without due process, places the whole faculty by implication in a status of vassalage.

The president of the *Harvard Crimson*, Jim Cramer, also had qualms about his paper's earlier actions. At the start of the next school year, in a full-page article based on interviews with Dean Ebert and with me, he expressed the conviction that I was sincere, though insensitive to the feelings of the students, and that liberals must answer my arguments with better justification for keeping up the scale of minorities admissions.

For several months after the public controversy colleagues at school seemed to avoid contact—I suspect out of embarrassment more than hostility. But after the atmosphere cooled off I found myself in a forgiving mood. This is expressed in the following comment (*Commentary* [August 1978]:70) on an excellent article in which a psychologist at the University of Michigan had castigated the academic community for allowing the affirmative action issue to undermine the ideals of honesty and objectivity.

> To the Editor of *Commentary:* Joseph Adelson ["Living with Quotas," May] has highlighted an important but little noted aspect of the problem of special minority admissions in our universities: the widespread unwillingness of university administrators to provide, even to their own faculties, the information that would be necessary for evaluating and improving these programs. Never has a free society coupled such a large experiment in education with such absence of feedback on the results. With academic leaders setting this example, no wonder an individual who dares to offer criticisms of methods in this area runs the risk of being labeled an enemy of the very goal of racial justice.
>
> But in seeking the cause of this tragic situation Mr. Adelson assigns responsibility in rather too personal terms. For example, in calling attention to the violent reaction at Harvard Medical School to an editorial of mine, which cited a particularly egregious instance of the lowering of standards in medical education, he ascribes the public censuring of my action to "the more thuggish elements" of the faculty. But it is hardly useful to invoke character defects, or to decry the lack of heroism: all of the several dozen faculty members in certain official positions reacted in this way to a highly distorted image in the media (though many privately shared my real views), and we must assume that such a large number of individuals are reasonably representative of the academic community. Rather, the lesson is that open and honest discussion of these issues in universities has become virtually outlawed by an atmosphere of extraordinary intimidation, combined with (or rationalized as) compassion and guilt, and compounded by the propensity of the media to maximize polarization.
>
> It is unlikely that universities can solve this problem by themselves. For while minority programs have brought in many students of whom we can be proud, the quotas encouraged by government bureaucracies and courts have also produced many problems, including the creation of a class of beneficiaries who oppose any alteration of this approach. This development is turning an originally moral issue into one of political power, before which academic institutions are particularly helpless. Unless the government comes out clearly against quotas, it will be difficult for many academic institutions to reverse this process and to regain their integrity.

Dr. Adelson's reply to this letter, in the same issue, included the following:

> Dr. Davis wrote the most circumspect editorial imaginable on a troubling situation at the Harvard Medical School. Despite his sobriety and caution, he was picketed, threatened with censure, denounced as a racist and fascist. Some of the Harvard administrative hierarchy joined in the abuse, including his dean and the president's office. It was clear enough that rule-or-ruin tactics were being applied: *we* will do whatever we please; we will admit as we please and grade as we please and graduate as we please; and if you dare oppose us, however gently, we will destroy your reputation. A message was being sent, not only to Dr. Davis, but to the rest of us as well. That is why the incident is important, so much so that at least three national journals of opinion have now published articles discussing it.
>
> I much admire Dr. Davis's generosity of spirit in seeking to exculpate his colleagues by framing the issue in larger terms; he is far more charitable to them than they were to him. But I think his analysis is only partially correct. Character *is* important. Even more important are the standards of civility governing conduct on difficult issues, standards which are to some considerable degree set and sustained by a university's leadership. Harvard does not now have that quality of leadership, to judge by its deplorable behavior in the Davis case.

One regret is that at the time of the storm I did not succeed, despite efforts through various intermediaries, in meeting with black students to give them a chance to probe into my beliefs and motives. However, a year later a letter in the *Black Health Organization Newsletter* at Harvard Medical School discussed a letter of mine in the *New York Times*, in which I supported the stretching of admissions criteria while opposing quotas, and the student's letter concluded:

> Whether or not Bernard D. Davis is a racist, one thing remains clear: his concern is for the production of top-notch physicians. And his letter to the *New York Times* may represent his attempt to convey this idea in a less controversial manner.

I was also gratified that several years later an excellent black scientist, Dr. Kenneth Olden, who had been a postdoctoral fellow at Harvard at the time of the storm, appointed me to the scientific advisory board of the Cancer Institute at Howard University when he became its research director.

A curious epilogue arose a year after the storm, when the curator of the Niemann Fellows at Harvard University, James Thompson, invited me to discuss with his group the role of the media in this episode. The Niemann Fellows are a selected group of journalists who are invited to spend a year at Harvard broadening their background. I accepted enthusiastically, expecting Mr. Thompson to lead a searching discussion of how the news media might improve their handling of such topics. I had also expected the journalists, knowing their field from the inside, to be willing to recognize weaknesses in the performance of their colleagues, and to help me to understand their problems, just as I would try to do if we were discussing problems in medicine.

On the day of the meeting Mr. Thompson phoned to say that he could not be present; but he did not postpone what he might have foreseen as a difficult discussion. Instead, he arranged for it to proceed under the direction of one of the fellows. The meeting was a shambles. From the first moment, I faced a hostile audience, and if any of the fellows were sympathetic with my view of the problem they were silent. One Hispanic reporter could not stand hearing me defend my position, and after ten minutes he slammed the door and left. I ended up with the curious impression that these presumably cynical reporters were actually naive enough to believe what they read in the papers. (The experience also reminded me that I had still not learned how to deal with the press.)

Conclusions

I have often asked myself how I would act if I could start with this problem all over again. I still believe my editorial is a reasonable statement, and I would not change it. I also feel that it was appropriate to bring into the open a serious issue, which was being fudged over on a large scale. However, I would now be much more circumspect in my replies to reporters. The theoretical obligation of the professor to be utterly honest in grappling with intellectual issues in the world of teaching is not good training for interacting with the world of politics and the media, and I suspect that I have been more naive than most of my colleagues in my failure to distinguish between these two worlds.

I have often also been asked why I took it on myself to call attention to extremely touchy problems that were well known to a great many people, and that our society was bound to correct eventually. The best answer I could give was the story of the old lady, gently awakened by an usher closing up a theater, who looked around at the empty seats and said "This is what comes of my following the theory that if everyone waited for everyone else there would be no rushing."

I do not question Dean Ebert's statement that his purpose in attacking my views so vigorously was to support the injured black students and to defend the school's program, and not to cause me any harm. I have always found him a kind person, though too willing, as an administrator, to accede to pressures. Tragically, however, the fundamental flaws in quota-based affirmative action programs have required innumerable rationalizations, and these have tempted many decent people, including the dean, into actions that undermine their values and their institutions.

The attacks orchestrated by the dean did not alter my official position at the school, nor did they interfere with my scientific work. Nevertheless, they no doubt lessened the impact of my subsequent writings on social problems, and perhaps also of my teaching. Moreover, the dean won the battle in the sense that I abandoned my role as an outspoken critic in school affairs. I also resolved not to write any more on affirmative action. However, a year later, when the *Bakke* case evoked extensive public discussion of preferential medical school admissions, I felt obligated to point out a possible solution (which appears in this part of the volume) that was being overlooked.

I am pleased to have the impression that the minority students in Harvard Medical School now are performing better on the whole than those of a decade ago. But this is only an impression. The faculty still has received very little objective data on comparative performance on the National Board Examinations, and no statistics on the records and career choices of the minority graduates. Meanwhile we devote one-fifth of our educational effort to this program.

A few years after the storm described in this article, and under a new dean, the Director of Admissions, Dr. Oglesby Paul, recommended that the autonomous Minorities Admissions Subcommittee be abolished, because it was using unsatisfactory criteria and was rejecting some of the best candidates. However, protests from black students and sympathetic classmates, supported by lawyers from outside organizations, raised such a storm (including a rally of two hundred students outside the doors at a faculty meeting) that the administration withdrew the proposal—although it later took steps designed to limit the autonomy of the subcommittee somewhat. What had begun as a program based on deep moral conviction, and intended to prime the pump and provide role models, had acquired a life of its own, frankly political and resistant to any reevaluation.

Nevertheless, we must eventually face the question of how long to continue programs of special treatment, and on what grounds, in medicine and also in other kinds of educational institutions. In medicine we would greatly strengthen the justification for continuation if we could demonstrate that a large fraction of the graduates do indeed serve primarily in minority communities. Not only is it important that these communities have access to physicians, but it is legitimate to try to satisfy a preference for physicians of one's own cultural background—provided they are competent. But if the beneficiaries of this special consideration in medical education are not filling this need we must ask whether or not the other gains for society outweigh the obvious injustices, to several groups. These include those better-qualified candidates who are bypassed, those minority candidates whose well-earned diplomas are devalued, and the patients.

Meanwhile, in any programs of special treatment we still have the problem, despite the *Bakke* decision, of choosing between stretching standards and imposing de facto quotas. If our universities and medical schools cannot find the courage to oppose the inroads of quotas on their mission of identifying and cultivating talent, they are betraying their trust. Quotas were introduced—and many feel that they were necessary—to open gates that were long closed to minorities; and it is essential to hold them open. But if, in carrying out this task, we force through these gates individuals who are poorly qualified we undermine both justice and effectiveness in our society, we impair the self-respect and motivation of those who receive handouts, and we even risk a backlash.

24

Minority Admissions: A Third Opinion

A year after my editorial on standards in medical education appeared the Bakke case reached the United States Supreme Court and provoked wide debate over affirmative action. I was disturbed to find all of the public discussion polarized in one or the other of two extreme positions. The American Jewish Committee and many other groups supported the claim of the plaintiff that admissions should be strictly colorblind. On the other side, much of the press, and a group of university administrations (including my own), defended de facto quotas, though in the circumlocutory terms that have generally characterized this debate.

I kept waiting to see whether someone might not suggest a possible compromise position: to increase the number of minority physicians by stretching standards to some limited, professionally acceptable degree, but not to set up quotas that would have to disregard such limits. When months of heated public discussion failed to produce this proposal I finally put it forward. Needless to say, it was gratifying that the decision by the Court produced a related compromise, forbidding quotas, but permitting schools to take into account past deprivation (though not race per se).

I would like to add an afterthought on this case. While quotas are unfair to those better-qualified candidates who are rejected, and to those minority students who do meet normal standards, the greatest unfairness is to the patients. They are entitled to have, and medical schools are obligated to produce, the best possible doctors (taking into account the variety of medical needs of the community). It seems unfortunate that the litigation of the Bakke case had to be conducted in terms of the rights of competing applicants, for that approach drew attention away from this more fundamental right of society. I am not impressed by the rationaliza-

---

Wall Street Journal, August 23, 1977.

tion offered by medical school administrators: that we cannot be rigorous in judging candidates because we really do not know what makes a good doctor.

Allen Bakke has sued the University of California for rejecting his application to its medical school at Davis while accepting less qualified minority applicants.

This case, to be reviewed by the Supreme Court in October, is widely interpreted as a contest between two alternative approaches to improving educational and professional opportunities of certain disadvantaged groups: (1) active recruitment and remedial education, followed by application of uniform standards for admission; or (2) the use of quotas (or their verbal equivalents) to ensure specific numerical representation of these groups.

Yet there is a third option: to adjust, supplement or stretch the standards for judging minority group applicants, but only within limits compatible with truly satisfactory performance. These adjusted standards would then determine the numbers admitted, rather than vice versa.

This approach is based on the belief that we cannot rectify past racial injustices by simply eliminating discrimination and practicing equality of opportunity. Rather, we must offer some degree of compensation for past limitations of educational and cultural opportunity. Accordingly, many medical and other professional schools accept the need to adjust their admissions criteria for certain minority groups.

The admissions committee at Davis, however, did not settle for adjusted criteria. Faced with more than 2,500 applicants, it established a composite score for each applicant on the basis of college grades, scores in the national Medical College Admission Test (MCAT), letters of recommendation, and interviews.

But instead of simply assigning additional points for a disadvantaged background it set up a minority list, with 16 places to fill, and a majority list, with 84 places. The composite scores for the minority list were necessarily weighted on the basis of social criteria, and it is noteworthy that Mr. Bakke found himself bypassed in favor of several minority students who had lower scores even after this adjustment for social background.

The Bakke case brings to a head the question of how a medical school should balance its primary obligation to screen for the aptitude of its candidates, and to ensure the competence of its graduates, with a more recently recognized obligation to help rectify past injustices and to help meet long ignored needs of underprivileged groups.

The idea of taking social background into account is not a radical departure, for admissions committees have always considered many attributes besides demonstrated academic ability and scientific knowledge. These include character, personality, motivation, career plans, work experience, geographical origin, breadth of interests, and sometimes even a "proper" social background. Now disadvantaged background is highly relevant.

The crucial point, however, is that no one set of qualifications can compensate for gross deficiency in another. Although one need not be extraordinarily brilliant to be a good physician, one must have reasonable ability to assimilate large amounts of knowledge and to analyze complex problems.

We must recognize, then, that while admissions committees have always considered nonacademic criteria, they have ordinarily worked only within a reasonably demanding range of academic credentials, as the best available indices of intellectual ability. And however deserving a student may be in other respects, a cutoff point on these grounds is essential if a school is to meet its obligations to society.

With a quota system, however, there can be no protective cutoff point, and there can be no adjustment to the year-to-year fluctuations in a school's applicant pool.

To be sure, a system of admissions based on adjusted criteria rather than on a quota does not guarantee a good solution, since the criteria could be so adjusted that the result would be the same. In practice, however, a primary focus on criteria rather than on numbers should make it easier to maintain reasonable minimal standards. For however much an admissions officer may be concerned about the welfare of future patients, the voice of his conscience has no opportunity to be heard if he must fill an arbitrary quota.

To get an idea of the latitude of the quota system at Davis, our best index is the MCAT examination, the one uniform and objective item in the composite scores. Mr. Bakke was in the 90th percentile of the national pool, while the *mean* of the admitted minority students was somewhere below the 50th percentile. (Some admitted individuals were therefore presumably far below 50.)

In the light of these considerations the Supreme Court may well decide against the quota system, and hence for Mr. Bakke, while at the same time encouraging schools to take previous disadvantage into account. If so, the Court will then face a second important question: whether disadvantage must be measured only in terms of individual economic and educational background, or whether membership in a generally deprived race or ethnic group may itself be considered relevant.

The strict interpretation of the Constitution would require the former, a colorblind procedure, affirmed by the Supreme Court of California in the Bakke case. However, if the high Court reaffirms that part of the judgment its decision would have a most unfortunate effect, for it would cause some of the best minority candidates to be bypassed.

This prediction is not simply theoretical. It is also based on the experience of a few medical schools, in which a special minorities admission subcommittee has quietly adopted the policy of favoring minority applicants with impoverished backgrounds over much better qualified applicants with middle-class backgrounds. The arguments are that the academic performance of the former does not reveal their ability, they are more deserving of compensation, and they are more likely to serve a minority clientele.

But apart from the fallibility of guesses about undemonstrated ability, and the dubious validity of predictions of a return to the ghetto, most medical educators would surely agree that their schools have a particular obligation to select the academically most promising minority candidates, precisely because of the present limitations of the minority applicant pool.

It would be unfortunate if the Court failed to accept this principle, and the pragmatic consequences that follow: if we are to adjust criteria, and if we wish to produce the best possible minority physicians, we must take race into account.

A number of universities have submited *amicus curiae* briefs supporting the University of California, with the aim of defending the autonomy of universities. However, the judgment they seek likely would involve court approval of a quota system, and that approval would give the Department of Health, Education and Welfare a license to impose quotas on all schools.

Hence the interests of the universities, and of society at large, would be better served by a qualified victory for Bakke: a decision that would forbid quotas and would encourage the maintenance of academic and professional standards, but would also validate the right of schools to consider ethnic as well as socioeconomic factors.

Though such a decision would cause an immediate decrease in the numbers of minority candidates admitted to some professional schools, in the long run it would also be a boon to minority education. For it would remove the stigma, present under a quota system, whereby the many excellent products of minority recruitment programs find their credentials devalued because of the forced inclusion of poorly qualified individuals.

25

Socioeconomic Quotas and Medical Education

This letter to the editor discusses a proposal that would try to solve a real and complex problem in distributive justice by extending the steamroller approach of quotas to socioeconomic groups.

To the Editor:

The federal scholarships proposed by Senator [Edward] Kennedy are aimed primarily at improving the distribution of medical care. They could also improve its quality, since they could broaden the pool from which the best candidates might be selected. Unfortunately, however, the Senator seems to visualize a different role for these scholarships. Thus in his letter of March 31 he stated that "for every American who gains admission to medical school three are turned away. And they are all qualified."

This assertion seems to suggest that admission is more a political right than an earned responsibility. For anyone who has passed certain courses in any college can apply to any medical school. Acceptance of this criterion of "qualified," in place of the traditional "most qualified," would have an enormous impact on medical education. Every medical educator knows that intellectual competence (among other qualifications) varies widely among candidates. Moreover, advances in medical science increase its importance. Since consumers cannot judge this quality well, medical schools have a particularly deep moral responsibility.

The Senator further complains (March 21) that thirty-seven percent of medical students come from the economic top twelve percent. Superficially this looks like proof of a large inequity. And inequities do arise, from financial barriers, biased members of admissions committees, and human fallibility. But while we should try to eliminate these factors we cannot predict how much of the cited disparity they account for: the uneven social distribution of the relevant abilities is another factor, and possibly a large one. Academic performance is strongly correlated with socioeconomic class in our society, whatever be the reasons; and however painful that fact, no conscientious educator can ignore performance in assessing individual ability.

Socioeconomic quotas for medical school admissions may have populist appeal, but they would undermine our commitment to excellence, in a profession that is responsible for human lives. And incompetent practice is a serious part of our current medical problem, though it is less visible to the public than inadequate availability and excessive costs. Politicization of medical school admissions will not help solve the problem of quality.

26

Letters on Racism and Affirmative Action

In the following three letters to the New York Times *I comment on items in that newspaper involving some aspects of the affirmative action controversy that were not related to medical education.*

Racism: The Numbers Fallacy*

In the *Times* of December 29 an article on "Disciplining Students on a Racial Basis," with the subheading "Minority Students Are Sent Home More Often," pointed out that in Dallas the rates of suspension from school have been much higher for minority students than for whites. While the writer noted that this disparity could be due to social differences, the rest of the article gave the impression that institutional racism in the disciplining of students is a widespread phenomenon. Yet no evidence was cited that the criteria had in fact been different for the two groups. Instead, expert testimony was cited that black children and their parents would perceive the Dallas schools as discriminating, and this perception would have an effect on the students.

This emphasis on numbers is merely the latest example of an error in reasoning that continues to distort the struggle for social justice: the assumption that since discrimination results in unequal numbers, then unequal numbers are proof of discrimination. In fact, unequal numbers, whether in admission, in achievements, or in any other respect, serve as

**New York Times*, January 19, 1975. Copyright © 1975 by The New York Times Company. Reprinted by permission.

flags to identify areas of possible discrimination, but they cannot serve *per se* as evidence. Articles and headlines that focus on numbers perpetuate the fallacy, and they may interfere with sympathetic efforts to get at the root of the problem. There may indeed be discrimination in the Dallas schools, but it hardly need be invoked to explain a high incidence of disciplinary problems in a group that enters school with many disadvantages.

What of the alternative argument: that even if the numbers do not reflect real discrimination they are still important because of the way they are perceived by minority groups. This emphasis on perceptions, rather than on the underlying realities, may be even more dangerous, for it is not a correctable assumption of fact but is rather a direct value judgment. It would seem to lead to the conclusion that disciplinary measures must be distributed on the basis of racial parity, rather than on the basis of equal treatment of individuals. What would the effect be on teachers, who must deal not with group statistics but with individual behavior? On the attractiveness of the teaching profession? On the other students, who would see justice defined in terms of a double standard?

Defining equality in terms of simple counts is easier than getting at the roots of discrimination—but it is no substitute, and it may well wreck our schools.

Of Jonestown and the Search for Utopias*

A letter from my colleague Alvin Poussaint (December 3) defended the People's Temple as a response to the needs of alienated persons desperately searching for a utopian promise. While I share Dr. Poussaint's sympathy for the victims of discrimination and disadvantage, I do not see how we can yet judge how many of the followers of the Reverend Jim Jones had a realistic basis for their extreme alienation and how many were mentally disturbed.

But apart from this consideration, Dr. Poussaint's letter reflects with particular clarity an attitude, shared to some degree by all of us, that has caused much mischief in our society: willingness to judge a movement (or a policy) entirely by its stated good intentions and noble goals, rather than by a critical evaluation of its methods and its consequences.

Thus he concludes that "the humanitarian experiment itself was not a failure; the Reverend Jones was." Further, "We cannot fault the entire rank-and-file because of the acute psychosis of their leader." He closes with the hope that this episode will not turn us against "the legitimate demands of *bona fide* social activists who seek a more sane and just

---

*

society." This last message surely deserves support—but it is weakened when the criteria for legitimacy or sanity are so broad that they would include the kind of social activism seen in Jonestown.

The most important issue here is not the validity of Dr. Poussaint's analysis of this cult: It is the need to recognize the roles of both feeling and reason in our search for a just society. This search surely requires continued strong emotional commitment, and Dr. Poussaint performs a valuable service in reminding us of this need.

But we court disaster if we fail to couple this motive force with an equally strong commitment to try to assess methods and results with our heads as well as with our hearts. When community leaders uncritically support utopian promises—as in the earlier endorsement of Mr. Jones by some of our most distinguished statesmen—they encourage demagogues and madmen to set up cults that prey on the weak and the desperate. Even worse, they encourage the development of self-defeating or dangerous policies in the broader society.

If we are truly sympathetic with the plight of the underprivileged, we should not weaken support for their cause by linking it to the search for utopias. Only by seeking solutions built on reality can we help the large numbers of people who have every reason to feel hopeless and alienated.

Blacks, Jews, and Affirmative Action*

In a September 30 advertisement, Julius Lester, a black writer, denounced the recent insensitivity of leaders of the black community to the moral issue presented by anti-Semitism and eloquently contrasted this position with the moral strength of Martin Luther King's earlier leadership. He further suggested that the present position of black leaders is a reaction to Jewish opposition to affirmative action.

Because Mr. Lester's statement is profoundly correct, and courageous, I hesitate to criticize a minor point. But to avoid further misunderstanding, it is important to clarify terminology, and to recognize that what Jewish organizations have been opposing is not affirmative action but a particular set of methods for implementing it.

Affirmative action is a concept introduced in a series of presidential executive orders in the conviction that we cannot achieve racial justice by simply eliminating further discrimination: we must also take positive, affirmative steps to speed the reversal of the effects of earlier discrimination.

Moreover, it was clearly implied in these orders that these steps

---

*New York Times, October 13, 1979.

would consist of active recruitment and provision of special remedial education and training, so that members of disadvantaged groups would truly have equal opportunity to compete. Subsequently, however, political pressures and bureaucratic decisions led to reinterpretation of affirmative action in terms of reverse discrimination and quotas—concepts quite different from equal opportunity.

Identification of affirmative action with quotas not only does an injustice to those who support the former but oppose the latter; it also jeopardizes the goal by linking it to unacceptable methods. Polls have repeatedly shown that the vast majority of our citizens, and even a majority of blacks, are opposed to quotas; so is the Supreme Court's statesmanlike compromise on the *Bakke* case.

It would be a tragedy if hostility of black leaders to those who oppose quotas should weaken liberal support for affirmative action.

Part Five

Public Concern Over Science

27

Novel Pressures on the Advance of Science

The late 1960s and early 1970s saw the rapid growth of an antiscience movement, as part of a wider rebellion against our social institutions. This paper, in a symposium of the New York Academy of Sciences, comments on that movement. It appeared shortly before the controversy over recombinant DNA arose, and the atmosphere that it describes may help explain why that controversy became so intense.

For several centuries science has been respected and admired as a major force for progress in human welfare. It has expanded the horizons for exploration by man's intellect, and its technological applications have given us mastery over nature even beyond the dreams of Francis Bacon. Hence the recent eruption of widespread public criticism of science[1] has come as a shock to scientists. This disenchantment was apparently triggered, after a delay, by the threat of nuclear annihilation, and it was enlarged by the highly visible effects of industrial pollution. The attack has now been extended to biomedical research and particularly to genetics, which is thought to be moving toward an invasion of man's ultimate sanctum, his selfhood or soul.[2] But even though some of the roots of the disaffection are well grounded, there is obvious danger that the public will have difficulty in distinguishing real concerns from exaggerated ones, and in balancing costs against benefits. For example, the marvelous prospect of gene therapy has generated more apprehension than pride; and though it is still distant, it has aroused public anxiety as

Annals of the New York Academy of Sciences 265 (1976):193–205 .

though it were just around the corner.

Every concerned citizen is aware of this novel discontent with science, and I cannot claim any special insight into its origins. However, I will comment on what I see as four main causes: actual costs of technological advances, unfulfilled expectations, the impact of science on our philosophical and social ideas, and the public's increasing contact with science.

Costs of Technology

The most prominent and concrete charge against science is that the social costs of its technological applications are now greater than its benefits. In the early nineteenth century this argument was presented by the Luddites, to protect their jobs, and by some humanists on esthetic grounds. It then had limited appeal, but as the scale of technology has grown it has become harder and harder to dismiss. In three areas this growth has generated crises. First, the consequences of escalation of *military technology* have become intolerable: we have endowed all future generations with the threat of nuclear catastrophe, and we have degraded our moral values by developing weapons whose use cannot discriminate between combatant and civilian victims. Second, though *civilian technology* yields more benign products, we now see that these are increasingly acquired at the cost of irreversible damage to our environment: exhaustion of nonrenewable resources, destruction of much natural beauty, and accumulating pollution of our air, water, and food. Finally, the increased *life expectancy* that was created by improvements in sanitation, medicine, and agriculture has led, for lack of accompanying birth control, to a Malthusian population explosion; and since it seems impossible to slow the proliferation soon enough, civilization may retrogress to the point where survival is no longer taken for granted but dominates our morality.

Technology has created additional problems by causing major changes in our life styles, whose effects are less apocalyptic but may be serious. It is asserted that the pace of modern life exceeds our biological adaptability and hence has greatly increased the frequency of illnesses due to stress.[3] A related argument is that the depersonalizing, monotonous work patterns in industry represent a loss of freedoms; moreover, this loss is not really compensated for by our increased material satisfactions since most of these depend on artificially stimulated needs rather than on real ones.[4,5] Finally, technological advances not only have altered our social patterns but also continue to cause them to shift rapidly, and this speed further threatens our social stability: for example, it increases the tensions between generations and between cultures, and it causes vocational skills to become rapidly obsolete.

Responding to these charges, we cannot deny that science has opened Pandora's box: it has given our species the ability to extinguish itself, if we cannot match our power of social adaptation to our power of destruction on a global scale. Indeed, Max Born has even questioned whether man was fortunate in discovering the key to the box.[6] But if we close the box now, or kick it, we can do no more than momentarily and irrationally relieve our frustration. With the problems both of armament and of population, radical shifts in our social habits and in our political practices are desperately needed, but a decrease in scientific activities will not help to provide the solutions.

With respect to the other unpleasant consequences of technological advance that I have mentioned I see little prospect of a major shift in public attitude. We will continue to accept these as costs that are inevitable in principle, though subject to modification in detail now that we recognize them. In the area of health, for example, it would clearly be desirable to decrease the strains of modern life, but we would hardly wish to do so if it meant giving up the medical and related technological advances that have so strikingly increased life expectancy. Our grounds for criticizing the focus of our culture on an endless increase in material comforts, convenience, and conspicuous consumption are stronger. But pressure for greater efficiency seems almost as inevitable in cultural evolution as in organic evolution (or in the second law of thermodynamics). Hence, while it is easy to decry materialism on moral or esthetic grounds, I can hardly imagine that most of the population would voluntarily give up various benefits of material progress unless the price increased a great deal.

False Expectations

Another basis for disenchantment with science is its failure to provide the expected magic solutions for many problems. One source of this disillusion should be easy to eliminate: excessively zealous promotion by scientists or fund-raisers, promising a rapid breakthrough in the control of cancer or some other disease. A more intractable cause, however, is the frequent emergence of unforeseen costs or unforeseen secondary consequences of technology: the toxicity of thalidomide, or disastrous ecological effects of the Aswan dam, or even the effects of the automobile, television, or oral contraceptives on our mores and social patterns.

The scientific community must be concerned with preventable abuses, such as inadequate testing for toxicity of a new drug. Moreover, we are now much more aware of our responsibility for trying to assess the future secondary consequences of a technological innovation, and the federal government has set up an Office of Technology Assessment. However, it is also important for the public to understand that grave consequences often simply cannot be anticipated, or cannot be prevented, and can be dealt with only as they arise. For example, we have virtually eliminated the infectious diseases that carried off a quarter to half of all infants and children a scant century ago—but that advance is largely responsible for the population explosion. Who could have foreseen that such a benevolent application of science would eventually lead to tragedy in underdeveloped countries? And even if the scientific community had recognized this problem much earlier, could it have persuaded the world to accept the desperate need to accompany death control by birth control?

In response to these painful problems, romantics of the counterculture and the New Left have accused scientists of dereliction of duty in not personally preventing their discoveries from being put to bad use. But this criticism is based on two tacit and very questionable assumptions: that scientists could have the power of such control if they wished, and that good and bad are self-evident. In fact, such ethical decisions ultimately involve the whole public. The training, special knowledge, and close involvement of the scientist may give him a broader perspective and a more objective approach, may increase his ability to foresee and to analyze alternatives, and may give him special reason to be concerned. But the scientific community is already seen as a priestly caste because of its special knowledge and its special contribution to changing what *can* be done; and nothing would alienate the public more than to have scientists become a true priestly caste, with the power to decide what *may* or may not be done. This presumed obligation, in the primitive form proposed, thus seems false. However, the scientist does have a more modest obligation: to communicate his special knowledge and insights when he sees the public interest threatened. And where economic or other conflicts of interest (especially those of an employer) interfere, social devices must be found to encourage this broadened concept of professional responsibility and to protect the professional who sounds the alarm.[7]

But the most important class of false expectations is that based on a philosophical misconception: an exaggerated notion of the ability of science to solve social problems. With the success of the Baconian approach in astronomy and physics, then chemistry, then biology, it seemed reasonable to expect the same approach to be equally productive when applied to the problems of society. Accordingly, when people see the genetic code being cracked and men being put on the moon, while violence and social decay spread, scientists are accused of being selfishly interested in satisfying their curiosity rather than in focusing on our most pressing problems. But the assumption that scientists could solve these problems is based on a naive earlier view, often called scientism, that failed to recognize the fundamental distinction between empirical questions, concerned with the nature of the external world, and normative questions, concerned with moral values. In principle science can answer the first kind but not the second: as Hume first pointed out, and G. E. Moore elaborated, as "the naturalistic fallacy."

Increased Scale and Increased Communications

The increased scale of research is another cause of increased public concern. Not only is the impact on our lives larger, but since the public now contributes a large amount of tax money to scientific activities, it has justifiably become more interested in judging their value and their propriety. In addition, development of instantaneous, worldwide communication in the news media has probably also contributed to public disaffection with science. Sensationalism in the press is hardly new; but competition for the attention of nationwide television audiences has led to increase stridency in the treatment of mishaps, errors of judgment, or unforeseen consequences.

Questions of propriety and ethics in the area of medical research are particularly likely to grip the public, since lives are at stake in conspicuous and dramatic ways. Moreover, with the increased scale more people are serving as subjects, and more of the research is done by people who view themselves primarily as investigators rather than as physicians. Accordingly, in this area ethical problems, long entrusted to the medical profession, have now been taken up by other groups.

The new profession of medical ethicist, coming to the problems from outside medicine, has had mixed effects. Responsible medical ethicists have helped to define the problems more sharply, and to emphasize the need to seek new solutions; but by encouraging contributions by experts who have had no personal experience with the complex emotional and scientific issues of medical care and research, they have opened the field for shallow pronouncements by individuals with little qualification. And the media have paid much more attention to these views than to those of professionals who have had responsibility for patients. In addition, some public-interest lawyers have also become interested in these issues and have brought with them the adversary process, with medical investigators now in the dock.[8] This development has not elevated the level of the discussion, and it probably brings us closer to excessively restrictive legislation.

Public exposure has clearly contributed to the correction of defects in the system. The medical profession has now accepted the need to require informed consent of all research subjects, and the need for research review committees to institutionalize the responsibility for defining acceptable risks and to enforce restraints. In addition, a variety of interested parties are struggling with realistic and hard questions: how the risks should be allocated, and how those who suffer damage should be compensated. These discussions undoubtedly benefit from the contributions of appropriate individuals from outside the medical profession.

On the other hand, excessive public involvement can also lead to unrealistic demands for risk-free research. For example, the thalidomide tragedy involved real errors, and regulations to prevent their recurrence were soon established; but these regulations were probably too stringent, for they have made it exceedingly expensive to gain approval of a new drug. We cannot measure accurately the losses that have resulted, but we know that the rate of introduction and approval of new drugs has fallen precipitously.

Other aspects of medical research may be facing the same kind of overreaction that drug testing elicited, with the media playing a similar role. Thus, in Wiseman's "Primate," shown on public television without rebuttal, clever editing has converted a presumably objective documentary into gruesome propaganda against research on animals. And enormous publicity has been given to such projects as the use of patients with a fatal illness to test for a possible immune response to cancer cells, or the Tuskegee study of the natural history of untreated syphilis in a group of blacks. In both these studies the insensitivity of the investigators to the rights of the subjects is indefensible; but it is also not fair to promote an overreaction by creating an image of cruelty comparable to that of the infamous concentration camp doctors. And it is easy, in the court of public opinion, to take cheap shots at the moral standards of an earlier era, and at actions that failed to measure up to our present level of virtue. (How will our standards look fifty years from now?)

My impression of my colleagues in clinical research is that they are a responsible group, who function not simply as investigators but also as concerned physicians. But unfortunately, we simply do not have hard data to prove whether the Tuskegee study is a rare aberration in a generally well-running system, or an example of a widespread callousness. Meanwhile, because of the special emotional appeal of medical problems, there is real danger of hasty and excessively restrictive legislation, subjecting many areas to the paralysis that fetal research now faces.[9]

Impact of Science on Our Values

The criticisms noted thus far are responses to highly visible effects of technology (including medicine) on daily life. But science has also drastically affected our inner lives: in expanding our intellectual horizons, it has shaken the previously accepted foundations of our value systems, and it has diverted the attention of many of the best minds from problems of values. Though these effects may not be of great concern to the man in the street they have long disturbed many people in the humanities, and they are probably receiving more attention now, whether as an undercurrent or more explicitly, since more obvious harms from science have generated a shift in public attitude.

With respect to the impact of science on the direction of our interests: until the advent of modern science the proper study of mankind was man, and the only approaches available were historical and speculative. With the discovery of the richly rewarding scientific method, the distribution of intellectual effort shifted enormously, from problems of sensibility and morality to problems of objective analysis. However, a reversal now seems to be taking place. The current cultural revolution is in part an expression of the conviction that science is not solving these ancient problems of human values—indeed, as I have already noted, it cannot and should not be expected to do so. Interest in formal ethics has been renewed, and it is further stimulated by the recognition that our new technological powers create new choices and responsibilities, and hence new ethical problems. Moreover, with some moral issues new technologies have virtually forced a drastic change in our outlook. For example, our changing views on abortion can be largely traced to effects of technology, both direct (antibiotics that make the procedure safe) and indirect (overpopulation; broader education and communication, generating emphasis on equality and hence on women's rights).

If science has temporarily inhibited scholarly interest in moral issues, it has had a permanent and shattering impact on the traditional foundations of our morality. Until recently these lay in a set of religious myths, almost universally adopted in the West (but not in the East), involving the postulate of a transcendental Creator as the source of our spiritual guidance. But science has now replaced earlier supernatural and animistic explanations of the universe by a coherent set of impersonal mechanisms. Darwin showed that man's uniqueness arose by the evolution of an extraordinarily complex brain and dexterous hands, and not by special creation or by the insufflation of a spirit; and Freud initiated a rational approach to understanding the irrational elements in our behavior. These developments split the rock underlying Judeo-Christian morality[2]: the true was separated from the good and the beautiful; and despite valiant efforts of religious leaders to preserve the traditional framework by reinterpreting earlier statements in symbolic terms, we have lost much of the consensus on ethical ideals that we had shared (however imperfectly we had attained them). I share Monod's view[10] that the growing response to this loss, and, to the failure of science to provide a basis for a replacement, underlies much of the tragedy, anxiety, and rootlessness of the present age. Indeed, I would suggest that the recent revival of attacks on the teaching of evolution may represent more than merely the irrationality of a dwindling band of fundamentalists: it may also reflect a long-standing public sense of spiritual loss, no longer suppressed now that criticism of science has become respectable.

Paralleling a sense of loss is a sense of increasing guilt, for increased abilities are inevitably coupled with increased responsibility. As long as a fatal disease, or starvation after crop failure, or poverty is considered an act of God it may create only a sense of tragedy; but when we recognize it as something that we might have prevented, it creates a sense of guilt. Freud[11] has suggested that the greatest source of discontent in civilized societies is guilt, arising from the need for restraints; and the enlargement of this burden by science may be a major cause of resentment.

The disturbing impact of science was unfortunately accentuated by a distortion of one of its greatest discoveries. Spencer's "Social Darwinism" extrapolated Darwinian evolution to society prematurely and, by analogy, rationalized prevailing economic and social principles by emphasizing exclusively the survival value of competitive behavior.[12] Modern studies in sociobiology, however, have demonstrated the inadequacy of this view of evolution: it is now clear that for social species long-term evolutionary success also depends on altruism, under appropriate circumstances.[13] This expansion of our understanding does not pretend to prescribe a scientific ethics, but it does help us to recognize the limits of the range of what is viable. Evolutionary theory is thus beginning to provide support for the traditional aims of moralists, but with two differences: it builds on a naturalistic base, and it recognizes explicitly, rather than grudgingly, the value of a balance between altruism and self-interest.

Technological advance has also been an indirect source of the current egalitarian revolution. By shifting the proportion of people living in affluence or in poverty, and by making the differences more conspicuous (both within and between nations), it has obviously contributed to the recent tremendous increase in emphasis on individual rights (including those of the poor and of unprotected inmates of institutions), and to the accompanying general decrease of confidence in authorities. In addition, as medicine has become more technically complex, expensive, and effective, our country has belatedly recognized that medical care is as reasonable a right as education. Accordingly, the traditional practice of drafting the poor as research subjects in teaching hospitals, long justified as the price for the superior medical care provided in these institutions, is no longer considered reasonable or defensible. Moreover, confidence in the profession has been specifically eroded by resentment at the economics and the frequently inadequate quality and availability of medical care, and by the publicization of horror stories (discussed earlier). For these reasons people are increasingly reluctant to trust the assignment of experimental risk to the judgment of individual medical investigators. Some other basis, ensuring a more voluntary and equitable distribution of risk, must be found.

Finally, the egalitarian movement has been distorted by some of its advocates, who have adopted an extreme environmentalist, Lysenkoist approach to human nature and have linked the desire to level social inequalities (which are accessible to our control) with a need to deny biological inequalities (which we cannot control). The resulting romantic view discards, as "elitist," the principle of trying to match abilities with responsibilities. Science falls under this fire, for by this definition it is indeed elitist: its essence is discrimination, by objective standards, between valid conclusions and valueless opinions, and hence between those who do and those who do not really understand a scientific issue. And modern populism has encouraged overt expression of long-smoldering resentment and envy of the special "secret" knowledge, and the power, influence, and prestige of the members of this meritocratic club.

Comments

This is quite a litany of complaints. What can scientists say in response?

Among the comments that I have already made I would emphasize the following. (1) Scientists simply cannot solve some of the problems they have been expected to take on. (2) They should try to foresee and warn about the consequences of their discoveries, but they cannot be expected to foresee them all, and they can only respond to unexpected ones as they arise. Since some of the things scientists do are now seen to be dangerous, we must accept substantial public input in the regulation of science, especially in medical ethics. But in working with concerned representatives of the public we must emphasize that a search for the chimera of risk-free research could be paralytic, and that regulation should reinforce, and not replace by minute legalisms, a sense of professional responsibility. I would add that taking the initiative to police ourselves, as in the Berg Committee recommendations on the construction of DNA hybrids, may help to avoid external imposition of excessively severe restrictions.

But though I accept the need for greater public participation I must confess to deep apprehension at the probable course of events. The character of much recent discussion has resulted in formulating the problem as one of setting up committees to decide which kinds of research to allow and which to forbid. This view assumes that there are areas of research requiring censorship. To analyze this assumption we must distinguish between censorship based on moral condemnation of the methods used in the research and censorship based on fear of the knowledge that it may yield.

Civilized communities have always had rules, if only implicit, that set limits on research *methods:* in particular, rules that forbid cruelty or unwarranted risk to people or to animals. If we are now becoming more sensitive on this issue and are requiring better definitions and stricter enforcement of such rules, and if we are broadening the concept of harm to include damage to the environment, it is a good thing. But to forbid a kind of research because of fear of its *consequences* is another matter. For freedom of inquiry is closely related to freedom of speech and to freedom of the press. Its absence from the Bill of Rights may simply reflect the fact that scientific inquiry had not yet become a substantial social activity.[14] We abrogate such freedoms only reluctantly, and in the face of a clear and present danger. Indeed, the scientist, deeply impressed by the history of unpredictable benefits from innumerable discoveries, and aware of the mutual dependence of different salients in the advancing front of knowledge, sees freedom of inquiry as virtually an absolute. But the layman does not generally share this view: he is likely to be much more sensitive to the tangible dangers of doing a piece of research (i.e., harm to individuals) than to the intangible costs of not doing it (i.e., failure to acquire knowledge). Freeman Dyson[15] has recently put the problem clearly: "The costs of saying yes can be calculated and demonstrated in a style that is familiar and congenial to lawyers, whereas the costs of saying no are a matter of conjecture and have no legal standing. . . . We must try to establish processes of decision-making that give the costs of yes and no an equal voice."

I would further emphasize a distinction between biomedical technology, which aims at preventing and alleviating illness, and the kinds of technology that aim at bigger and better consumption. Problems of exhaustion of nonrenewable resources, and of unequal distribution among the world's peoples, may well lead us to decide eventually that we must curtail the latter kind of technology, since we cannot indefinitely expand (or perhaps even maintain) the high level of consumption that we have grown used to in the West. And such a development would surely lead to a slowing of the related basic research. But biomedical research poses no such threat of a cataclysm, except via overpopulation and that can, in principle, be controlled without interfering with medical advances.

NOTES AND REFERENCES

1. "Science and Its Public: The Changing Relationship," *Daedalus* (Summer 1974).
2. G. Stent, "The Dilemma of Science and Morals," *Genetics* 78 (1974):44–51.
3. R. J. Dubos, *Man Adapting* (New Haven, Conn.: Yale University Press, 1965).
4. L. Mumford, *The Myth of the Machine* (New York: Harcourt, Brace & World, 1967).
5. T. Roszak, *Where the Wasteland Ends* (Garden City, N.Y.: Doubleday-Anchor, 1973).
6. M. Born, *My Life and My Views* (New York: Scribners, 1968).
7. J. T. Edsall, "Scientific Freedom and Responsibility," *Science* 188 (1975): 687–693.
8. *Experiments and Research with Humans: Values in Conflict* (Washington, D.C.: National Academy of Sciences, 1975).
9. W. Gaylin and M. Lappe, *Atlantic Monthly* 66 (May 1975).
10. J. Monod, *Chance and Necessity* (New York: Knopf, 1971).
11. S. Freud, *Civilization and Its Discontents* (New York: Cape and Smith, 1930).
12. R. Hofstadter, *Social Darwinism in American Thought* (Philadelphia: University of Pennsylvania Press, 1944), published in paperback in 1955 by Beacon Press.
13. E. O. Wilson, *Sociobiology* (Cambridge, Mass.: Harvard University Press, 1975).
14. D. Stettin, "Freedom of Inquiry," *Science* 189 (1975):953.
15. F. J. Dyson, *Bulletin of Atomic Scientists* (June 1975): 23–27.

28

Fear of Progress in Biology

The antiscience movement, discussed in the preceding paper, is part of a broader problem: the decreasing confidence of Western society, or at least of its intellectual community, in the value of the kinds of progress that advances in science and technology have brought us. This essay was prepared for a conference on that problem sponsored by the American Academy of Arts and Sciences. It has a broader perspective than many of the pieces in this volume, weaving in a number of themes that are also treated in other, shorter pieces.

The Changing Contract between Science and Society

For centuries the scientific community enjoyed virtually complete autonomy in choosing the directions of its research, and also in regulating any attendant hazards; and the record seemed to be one of almost pure benefit and achievement. In recent years, however, we have seen increasing concern about where science is taking us, and increasing demands that the public determine what scientists may or may not do.

This new attitude arose in response to belated recognition that the technological applications of the physical sciences generate large social costs as well as benefits—costs that range from the threat of nuclear

---

In *Progress and Its Discontents*, ed. G. A. Almond, M. Chodorow, and R. H. Pearce (Berkeley, Calif.: University of California Press, 1982).

annihilation to despoliation of the environment. The biological sciences at first seemed immune, since their major applications—increased control over disease and increased food production—are so obviously humanitarian. To be sure, these successes have also created problems: the resulting population explosion may turn out to be the greatest underlying cause of social unrest, and the development of very expensive medical procedures raises problems of distributive justice. But these issues, though present, do not yet seem to loom large in current public apprehension over biomedical research. Instead, concern over possible or hypothetical future dangers has been leading to demands for control over the basic research itself rather than over its applications.

Before discussing these presumed dangers from biology, I would like to note briefly some of the more general reasons for the recent growth of disaffection with science. (1) The rapid advances in many areas of science have caused even the most improbable future projections to be taken seriously. Accordingly, scenarios belonging in science fiction become sources of anxiety—especially in biology. (2) Many short-term benefits of technology have turned out to have long-term costs, and the scale is growing. Technological advance in general has therefore become suspect. (3) Because of the success of science and technology in reducing many of mankind's traditional ills and hazards, expectations of absolute security have replaced a mature recognition that costs generally accompany benefits. (4) The important distinction between science and technology is often blurred. For example, even pure biology is tainted by the use of defoliants in Vietnam. (5) Since science is inherently elitist (in a sense depending on ability and achievement rather than on social origin), the egalitarian thrust of our era has created guilt among many scientists and has weakened their confidence in the moral status of their enterprise. This development, and the increasing dependence of research on public funds, has encouraged acceptance of the neo-Marxist view that science is primarily an instrument of the prevailing political system rather than a methodology for seeking universal, objective truths about nature. (6) Major failures of our political institutions, often linked to advice from academic experts, have led to a general mistrust of institutions and experts. (7) As science and technology become more complex, they influence the life of the ordinary citizen more, while at the same time he understands them less. The disparity, as well as the speed of the resulting changes in our way of life, generate uneasiness. (8) When scientists hold conflicting views the mass media find it hard to assess their judgment and credentials, and the more sensational claims of hazard are likely to be featured. (9) The cohort of activist students of the 1960s has now reached influential positions in the media, and also in science. In particular, such groups as Science for the People have chosen genetics, rather than our political and economic structure, as their focus, and they have acquired attention out of proportion to their numbers. (10) The program of the Enlightenment has failed: the relative freedom from want created by technology, and the spread of rationality encouraged by science, have not resulted in general moral progress. In fact, science has undoubtedly contributed to a weakening of the moral order by undermining the traditional supernatural foundation for a moral consensus without providing an alternative. Moreover, while the uncompromising emphasis of science on objectivity has provided great intellectual strength, the price has been a shift of much of our intellectual focus away from subjective values. (11) The Judeo-Christian assumption of man's right to unlimited multiplication and to unlimited dominion over nature arose at a time when the spread of agriculture encouraged an increase in population. The present need to reevaluate this assumption, in the light of diminishing resources, adds another dimension to our sense of moral crisis.

On all these grounds, changing public attitudes could lead to a real contraction in the support and the prestige of science. Such a development would seem sad to those of us who still see science as a source of major benefits to society: power to improve our physical conditions and our security, deeper understanding of ourselves, and delight in the expression of man's intelligence and creativity.

We might note that Gunther Stent has also predicted a contraction in science, but on quite different grounds: the exhaustion, quite soon, of the possibilities for further interesting scientific progress. He suggests that we will then replace the Faustian striving for personal accomplishments and for increased control over nature by the Taoist goal of a static, harmonious adjustment to nature.[1] Elsewhere in this volume John Edsall has considered this possible limitation to future progress in biology. I would simply like to add my doubts that the end of an exciting age of fruitful exploration could lead smoothly to a comfortable golden age of enjoying the fruits. Instead, a dense world population, competing increasingly for dwindling and unevenly distributed resources, seems more likely to drift desperately into a flight from science and rationality, and hence into a new Dark Age.

Against this rather discouraging background, I shall try to assess the reality of the assumptions underlying three widespread fears about advancing knowledge in genetics: that these advances will create dangerous products, dangerous powers, and dangerous insights into human nature. I shall close by discussing the role of objective knowledge in our intellectual life, emphasizing not only its value but also its limits.

Possible Dangers from Genetics

Dangerous Products

Though microbiologists have been cultivating pathogenic organisms for a century, public concern about possibly dangerous biological materials did not become widespread until the recent development of the recombinant DNA methodology. In this technique a segment of DNA from any source can be spliced into a DNA molecule in the test tube and then replicated (cloned) in bacteria. It has thus become possible to isolate any gene in quantity, to study its function in a simplified environment, and to manufacture many desired products. A decade ago such a discovery would have been greeted solely as a remarkable breakthrough. In the current atmosphere, however, public discussion has focused much more on the risk of inadvertently creating and releasing dangerous new organisms.

This contingency was initially raised by a group of molecular biologists. Their concern was very much in the tradition of responsible science. But they departed from tradition in one respect: perhaps in order to disprove the recent charge that scientists have been elitist in making decisions for the public, they expressed their concern publicly before they had time to explore the matter extensively. Their candor was acclaimed initially, but it soon gave rise to widespread public anxiety, particularly after a handful of other scientists raised an alarm.

By now much of the apprehension has subsided. It may be of interest to summarize briefly the main scientific reasons, which I have reviewed elsewhere in greater detail. (1) After several years of work, in hundreds of laboratories, with such chimeric bacteria, the hazards have remained entirely conjectural: no illness or environmental damage has been traced to this source. (2) Mutant bacterial strains have been developed with a remarkable, novel safety feature: they require special nutrients that are lacking outside the laboratory, and without these compounds the cells rapidly self-destruct. (3) It has recently become clear that bacteria transfer DNA from one species to another promiscuously (employing, in fact, the same enzymes that investigators extract and use for in vitro recombination). This finding makes it extremely likely that the recombinants with human DNA now being made in the laboratory are not a novel class of organisms after all, since *E. coli* in the mammalian gut would occasionally take up DNA released from dying host cells (as well as DNA from other bacterial cells). (4) In nature novel mutants are continually being generated, and only an infinitesimal fraction of these innovations pass through the sieve of natural selection and survive. Moreover, this survival depends not on the properties of a single gene but on the adaptive value of a balanced set of genes; and insertion of DNA from a distant source, in the new technique, is almost certain to impair that balance. (5) Since this insertion adds only about 0.1 percent to the DNA of the host *E. coli*, a recombinant will retain the mode of spread of *E. coli* and will be restricted to the habitat of that organism (the vertebrate gut). Hence epidemiological experience with pathogens closely related to *E. coli* is pertinent—and it is reassuring. Indeed, from the inception of the debate no expert in epidemiology or infectious disease supported the view that *E. coli* might inadvertently yield recombinants as hazardous as the already known major bacterial pathogens—organisms already selected in evolution for the ability to spread, and likely to turn up in any diagnostic laboratory at any time. Moreover, though the history of microbiology includes several thousand laboratory infections, and a few microepidemics, no large epidemic of any pathogen has ever arisen from a laboratory. (6) Ironically, views on the possible spread of tumor virus genes by bacteria, which started the discussion, have rotated 180 degrees. Viral DNA cloned in bacteria, from which it can be released only as naked DNA, is over a million times *less* infectious to an animal than the same DNA released from its natural animal cell host, as a complete viral particle with a protective coat.[2] Hence, an investigator can now prepare such DNA more safely by cloning it in bacteria than by the conventional (and unregulated) methods in animal cells. (7) Since mild pathogens are much more common than severe ones (for example, the common cold versus the influenza virus), it seems exceedingly unlikely that a serious pathogen could be inadvertently produced before milder ones had appeared and warned us.

With the recognition of these facts, and after an enormous amount of discussion, public anxiety abated. The very real threat of restrictive and even punitive legislation has been dropped, and the National Institutes of Health guidelines regulating this research have gone through two successive stages of relaxation. But reason prevailed only after a great deal of time and money had been spent fighting exaggerated or nonexistent dangers. Moreover, a large regulatory bureaucracy was set up. Starting on this slippery slope may be the greatest price of all, unless the experience helps us to develop better mechanisms for evaluating risks in highly technical areas in the future. For such bureaucracies not only are costly in time and money, and occasionally obstructive: invasion by their rigidity also inhibits the sense of playfulness and of artistic creativity that has characterized much of the best scientific research.

The Need for Improved Assessment of Possibly Dangerous Actions

Concern over actions that might create dangerous materials is clearly legitimate in principle. Moreover, in practice scientists have had little trouble in agreeing with public agencies on regulations over demonstrably dangerous materials—inflammable, explosive, toxic, or radioactive. Problems arise, however, when the hazards are matters of judgment, rather than of demonstrable fact. In both circumstances the assessment of the hazards is a technical job, best handled by those with the requisite special knowledge, while the subsequent process of making policy should involve a wider group.

A move in this direction would require an adjustment of attitudes in both the general community and the scientific community. On the one side, the record justifies a restoration of public trust in the sense of responsibility of the scientific community when it is asked to provide objective and informed judgments on the technical matters in which its members have special knowledge. At the same time, having recognized that science and technology present hazards as well as benefits, we cannot go back to an earlier era in which scientists were trusted to make all decisions that involved science. Since risks and benefits are generally noncommensurable and are unevenly distributed in society, and balancing them involves not simply technical questions but also value judgments, the general community must be involved in the later stages of decision making.

In asking for trust, scientists must also recognize a new responsibility toward the public. With the growing impact of technology on society, and with large economic stakes biasing many sources of information, we need greater watchfulness and initiative on the part of those scientists closest to a new development—willingness not only to provide answers when asked but also to expose abuses as soon as they become identifiable. In addition, the scientific community must not pretend to more expertise or more objective knowledge than it has, especially in those fields where the knowledge is diluted by a great deal of uncertainty.

It is not obvious how confidence in the scientific community can be restored. In a broad sense education is perhaps the only way. But I would suggest that for this purpose an informed public is not simply one that is exposed to news about recent discoveries. Much more important, and more difficult, is education on the nature of scientific activity. Legislators, in particular, need to know that discovery is inherently unpredictable, and not purchasable quite like a commodity. Another point, particularly important in asking for trust, is that scientists are intensely trained to be honest in handling their data—not because they are more virtuous than other people but because their findings are valueless unless verifiable, and they know that nature has the last word. Finally, though scientists are not infallible, and though at the growing points of science interpretations rise and fall, the scientific community has evolved extraordinarily effective communal mechanisms for discriminating between true and false conclusions. And that community sometimes needs quite a bit of time to digest and exchange information before a conclusion is firm enough to warrant public attention.

Perhaps we can learn a lesson from the reaction to recombinant DNA, which retrospectively seems to have been close to hysteria. The record of the scientific community was certainly a highly responsible one. Indeed, the problem arose because the molecular biologists who created the novel techniques were carried away excessively by concern over the theoretical possibility of creating novel epidemics, and it took time for the highly reassuring information from epidemiologists and evolutionists to have an impact. If the matter could have remained within the scientific community in the first state of a two-stage process, we would have avoided a futile and expensive exercise—expensive in terms of time, money, public anxiety, and the morale of the scientific community.

Dangerous Powers from Genetics

Let us now shift from research that may yield dangerous products to research whose results may give us dangerous powers—a problem that raises quite different issues of social policy. In the biomedical sciences the powers most feared are those of genetic engineering, that is, directing changes in the genome of an organism.

In medicine the phrase genetic engineering, with its cold overtones, does not seem very apt, for the goal is simply gene therapy: supplying the single genes whose absence in normal form cause various hereditary diseases. By itself such an intervention would surely be as legitimate as the daily replacement of a gene product, such as insulin. Nevertheless, the idea has generated alarm. Some critics see such manipulation as something akin to sacrilege—the invasion of sacred territory. But most of the concern no doubt has a more pragmatic basis: fear that if we develop such techniques for medical purposes, those in power may employ them for political purposes—not to cure or to prevent diseases but to manipulate personalities. This belief, as an undercurrent, clearly added to public anxiety in the recombinant DNA debate.

If I believed such political applications of genetics were at all likely, I would share this apprehension. But just as with the dangers from recombinant DNA, I find the technical facts highly reassuring, on several grounds. First, even therapy of single-gene defects still seems far off (except for the precursors of circulating blood cells, which are so loosely organized in the bone marrow that they could conceivably be replaced by other cells). In addition, even if gene therapy for monogenic diseases should be achieved, the complexity of the genetic contribution to individual differences in behavior presents a huge technical obstacle to its manipulation. For though we know virtually nothing about these genes, we can be sure that their number must be very large. Intelligence, or altruism, or any other behavioral trait is polygenic: that is, it is not determined by a single gene but instead is influenced by many genes, interacting with each other and with the environment. The problem of identifying such a large, coordinated set of genes, and replacing them in a predictable way, is very much greater than that of single-gene therapy.

Still another obstacle to genetic manipulation of personalities arises from the fact that the function of the brain depends on an intricate network of specific cell-to-cell contacts, and most behavioral genes act by guiding the development of that circuitry. Since these genes will have done their work before birth, gene transfer could not conceivably rewire an already developed brain. In principle, one could circumvent this difficulty by replacing genes in germ cells. But this procedure not only would be technically even more difficult: it would also be useless, for one would be investing great effort to change some genes in a germ cell whose other genes were still an unknown, chance combination.

Social factors further limit the possibility of genetic control of behavior. Though it has been suggested that genes might be manipulated secretly (for example, by dissemination in a virus), such fantasies cannot be taken seriously, if only because of the complexity of the complement of behavioral genes. We must therefore assume that genetic manipulation would require cooperation of the subjects: and any population willing to cooperate in this way would already have lost its freedom. Moreover, genetic manipulation of personalities, if ever feasible, would have to compete with other, less elaborate, and less costly means that are already at hand or in process. These include the familiar psychological methods (amplified by modern methods of mass communication) and the methods of pharmacology, neurosurgery, and even eugenics (that is, selective breeding for the desired traits).

For these many reasons discussions of ethical aspects of genetic intervention—recently a major topic in biomedical ethics—may be seen as theoretical exercises in moral philosophy rather than as analyses of present or imminent social problems. Indeed, that so many scholars have taken the issue seriously testifies to the penetration of science fiction into the contemporary image of science. The alarm may abate as the issue is clarified and the anticipated powers remain remote. On the other hand, because of their emotional appeal the attacks on genetic engineering may threaten the highly desirable medical goal of gene therapy, just as similar attacks have effectively discouraged research on chromosomal aberrations in infants.[3]

Unlike gene replacement, another type of genetic manipulation of humans has seemed quite close at hand: the creation of genetic copies of an individual. Such cloning has been successfully accomplished with frogs, by implanting nuclei from somatic cells of an embryo into egg cells. Ten years ago it seemed self-evident that improvements in technique would sooner or later extend the procedure to mammals; and this scientific advance, of obvious value in agriculture, would create serious moral problems if it should be extended to man. However, the prospect has now changed, with recent indications—though not yet decisive evidence—that various fully differentiated cells do not have quite the same genetic information as the embryonic cells. Hence cloning of mammalian adults may well be unachievable, for fundamental reasons rather than for reasons that might be overcome by advances in technique. If so, human cloning by nuclear transplant, aimed at copying individuals with already demonstrated traits, loses its potential interest—and its threat.

Should We Ever Restrict Knowledge That Might Yield Dangerous Powers?

I suggested earlier that in principle restrictions on research procedures involving potentially dangerous materials are clearly legitimate: the problem is how best to go about the job of assessing hazards when they are uncertain. However, when we consider research that might give us dangerous powers we face a more fundamental question: not how to improve procedures for setting limits, but where the limits should be set. Should we limit the search for certain kinds of knowledge, or only limit its applications? Waving the flag of Galileo may no longer be an adequate answer. For even though open societies have a long tradition of defending free inquiry as a mode of free expression, we find serious people today suggesting that science has reached a stage where it might yield powers too hot to handle.

We cannot exclude this possibility logically. But we can exclude restrictive actions on other grounds: that we cannot identify such undesirable knowledge in advance. Being able to foresee a conceivable dreadful application is not enough. All knowledge is double-edged; and to justify proscribing any knowledge, we should be able to provide convincing evidence that the probable peril outweighs the probable gain.[4] No basic scientific knowledge has yet met this test.[5] Indeed, it is difficult to see how any knowledge could meet the test, for we simply cannot foresee all the applications of any knowledge; even less can we foresee all the social consequences of these applications. It would follow that we can still best serve society not by blocking any particular knowledge but by better controlling its applications. We should therefore seek to improve our methods for recognizing early the costs and dangers, as well as the benefits, of various applications, and we should resist economic and other pressures for automatically proceeding with all possible applications (the technological imperative).

Knowledge Believed to Endanger Social Justice

Let us now consider the third concern over advances in genetics: the production of knowledge that would undermine the foundations of public morality. This old source of fear of scientific progress arose with the heliocentric theory and reappeared in the reaction to Darwin. After the Scopes trial it seems unlikely that this form of antiscience would remain a matter of concern. But today we see not only interference by creationists with the teaching of biology in public schools. A much more serious problem is an ideological attack, spearheaded within scientific circles, on the study of the biological roots of our behavior, and particularly on the study of human genetic diversity. This subject cuts even closer to the bone than did the earlier question of man's origin, for the results may conflict directly with assumptions about human nature that are difficult to question because they underlie strong political convictions. Moreover, a restriction on presumably dangerous insights is not simply a problem for biologists: it raises the question of intellectual freedom for the whole scholarly community.

It is ironic that human implications of evolution and genetics are now opposed primarily from the left. Darwin, in contrast, was seen as a threat not by liberals but primarily by religious traditionalists. Moreover, in the 1930s the distinguished British geneticist and Marxist J.B.S. Haldane could strenuously oppose Hitler's pseudogenetics and at the same time emphasize that real study of human behavioral genetics offers great promise for education.[6] More recently, the sad fate of genetics in the Soviet Union under Lysenko has offered a vivid warning against subordinating the search for objective knowledge to ideology. Nevertheless, the current attack has evoked wide sympathy, and it has created an atmosphere of intimidation: few graduate students today are likely to enter the field of human behavioral genetics.

The reasons for fear of this field are evident. One is the widespread conviction today that genetic differences between people, however real, should not be discussed in public lest they discourage or limit egalitarian aims. Older reasons are the past history of political misuses of genetics, and the simplistic extrapolation from early evolutionary concepts to Social Darwinism and the use of pseudogenetics to support the racism of Nazis and white supremacists. Given this tragic history, we must recognize that genetics could indeed again be misused to rationalize discriminatory practices; and we should be especially concerned about this possibility today, when we have finally begun to rectify our legacy of race discrimination.

But though this history has led to the assumption that studies of evolutionary and genetic aspects of human behavior are bound to have a reactionary social impact, this conclusion does not follow. For if we look closely at the past abuses of this area of science, we will find that in each case the politics has distorted the science, rather than being derived from it. Paradoxically, then the ideologically oriented critics of behavioral genetics today are the true spiritual heirs of the tradition that they appear to be opposing. For these critics would again subordinate scientific knowledge, though in a different direction, to social preconceptions, while the proponents of genetic inquiry are defending the universality and objectivity of science against political undermining, from either the right or the left.

One might counter that some kinds of knowledge could threaten the goals and values of a just and decent society, and so it is callous for the scientist to seek knowledge, in the tradition of disinterested inquiry, without regard to its political consequences. But however humanitarian the intent of this criticism, it misconstrues the relation between knowledge and justice. Justice is a social construct, and it is constantly evolving as we adapt to changing circumstances. Moreover, scientific findings cannot specify a particular construct as the correct one. They can only test some of the assumptions about human nature that underlie our efforts to develop adaptive social institutions; and the degree of correspondence between these assumptions and reality strongly influences the success or failure of our social experiments. Since science may thus reveal, but does not create, the reality that plays this role, it is difficult to see how scientific knowledge itself can be a threat to justice. To be sure, in its social applications knowledge can also be distorted, or misunderstood, or prematurely extrapolated. But as with knowledge that creates double-edged powers, our problem is to avoid the abuses, not the possession of knowledge that creates insights.

This point can be illustrated more concretely by a deeper look at the history of the problem of racial justice. While genetics has been subject to very well-known past abuses in this area, as we have just noted, it has also made a positive (but unrecognized) contribution to our modern conception of racial justice. Specifically, the nature of race was reviewed for centuries in *typological*[7] terms, that is, on the basis of the Platonic view that any class of entities is best understood in terms of an ideal type, the concrete variations between individuals being of trivial importance compared with the essential characteristics of the type. This prescientific view led to the belief that a person's race defines his potentials, thus providing a rationalization for racism (that is, for discriminatory treatment of individuals on the basis of identification with a given race). However, this assumption has been demolished as modern evolutionary theory and population genetics have replaced the earlier, vague social notion of race by a precise biological concept. Biologists now define races not in typological terms but in statistical, *populational* terms, that is, as subpopulations, within any species, that have been separated for enough generations so that their total gene pools have evolved significant differences in gene (allele) frequencies. Moreover, human races are highly heterogeneous (that is, they consist of individuals with widely varying potentials), and they all overlap in their distributions.

I would suggest that this biological insight has made a major, hidden contribution to the modern revolution in our social attitude toward race. The fundamental arguments for eliminating racism are, of course, moral and political; but these arguments are built on a tacit understanding that the earlier, typological view of race is false. If that view still prevailed, it is doubtful that the moral arguments would be convincing. Again, our problem is not to avoid scientific knowledge in socially sensitive areas but to avoid unsound or misinterpreted knowledge.

Scientism and Objectivity

The Dangers of Scientism and the Value of Objectivity

Having discussed public concern over both real and conjectural dangers generated by science and technology, I shall now discuss another source of public disaffection with science: inflated claims for its power. For example, in the early, heady days of molecular genetics it was tempting to boast about future possibilities for reshaping man; but as the possibilities have seemed to be coming close, these boasts have now stirred up fears.

A much broader basis for exaggerated claims has been the view called "scientism": the expectation that the advance of science will ultimately provide definitive solutions for the problems of society.[8] Our failure to find these solutions leads to criticism of scientists for wasting their talents on the wrong problems.

It is quite understandable that scientistic predictions should have been widely accepted at an earlier time, for science seemed to be a universal problem-solving machine, with no inherent limits in its application to increasingly complex problems. Today, however, we recognize limits to its scope. One reason is the size of the gap in complexity between human social problems and problems in the related natural science disciplines (neurobiology, ethology, and evolutionary biology). Another is that in social processes small causes can have large effects.

But the most fundamental reason is epistemological: social actions involve value judgments, and in principle problems of values have no objectively demonstrable correct solution. Hence solving a problem in physics or in biology does not have the same meaning as solving a social problem. Indeed, in the absence of objective criteria, some would say that we can only try to manage, and should not speak of solving, social problems.

These limits, however, do not imply that science is irrelevant to social problems. On the contrary, it can help in several ways. First, through the tools provided by technology it can broaden our control over nature and hence our range of options. (Of course, increased options also create new problems.) More fundamentally, science is very good at predicting consequences of alternative actions—and such predictions enter, tacitly if not explicitly, in our selection of a value system, and in making concrete decisions within that system. And most directly, the development called sociobiology reflects a recognition that human social behavior has biological as well as cultural roots, and it encourages the hope that a deeper understanding of both will be helpful. As E.O. Wilson has pointed out, "the genes hold culture on a leash": our values are neither entirely arbitrary cultural constructs, on the one hand, nor rigidly determined products of our genes, on the other.

Nevertheless, the leash is long, and it encompasses a very broad range of possible social patterns. Sociobiology therefore cannot be expected to prescribe any particular pattern as correct for a given set of circumstances, nor can it tell us how we ought to balance conficting values. But it should be able to have an adjuvant role, helping us to incorporate a deeper knowledge of the realities of human nature—including its diversity as well as its universals—in deciding between alternative courses of action. It would be presumptuous to try to estimate how large this role will become, but meanwhile it seems important to avoid scientistic predictions, which can lead both to fear of their fulfillment and disappointment at their failure.[9]

The social sciences are the largest area where scientism has created disillusion. For despite the emphasis of these disciplines on objective, rigorous, and quantitative studies, in most areas their conclusions and predictions are far less certain than those in the natural sciences. Nevertheless, governmental and other social agencies have increasingly relied on these predictions as a basis for action, as though they were virtually infallible products of an all-powerful scientific method. The resulting disappointment has no doubt contributed to the antiscience movement, because the public (and even the National Academy of Sciences) closely identifies the social sciences with the natural sciences. Public education on the nature and the limits of science should therefore include a clear distinction between these two uses of the word "science."[10]

Soft-core Scientism: The Role of Objective Knowledge

Gunther Stent, long a stimulating commentator on the idea of progress, has recently put forth a more extended conception of scientism. He agrees with the view that science cannot provide authoritative, correct solutions to moral problems—an expectation that he calls "hard-core" scientism. However, he disagrees with the view, which I have defended above, that we should always welcome the "adjuvant" function of objective knowledge because solutions to moral problems will be more effective if they avoid contradictions with reality. He calls this view "soft-core" scientism: "the dubious empirical proposition that the realization of moral aims is necessarily impeded by acts which are motivated by objectively false beliefs."[11]

Stent cites two examples. In the first he concedes that the false belief of the Hopi Indians in the effectiveness of the rain dance may have harmed their agriculture, but he suggests that it may nevertheless have provided a greater benefit by promoting communal cohesion. His second example comes closer to home: the problem, already discussed above, of research on the hereditary basis of intelligence. Stent criticizes equally those who consider such research essential and those who consider it pernicious, because both accept the assumption that if there is genetic diversity, it ought to be taken into account in the organization of society. He proposes, instead, that we should ignore such diversity because the communal cohesion fostered by the false belief in innate human equality could outweigh the losses due to the resulting falsely based educational system.

But a myth can promote communal cohesion only if it is widely enough accepted by the community. The Hopis could all believe in the power of the rain dance because their world view would not lead them to test it. In a modern, science-based society, in contrast, any testable assumption or claim will inevitably be tested, and if it proves to be objectively false, it will be disputed. The expected communal cohesion then becomes dissension between rationalists and believers. (We need only note how Western religions have increasingly narrowed their jurisdiction, abandoning their earlier role of providing supernatural explanations in areas now taken over by science.) Accordingly, however convenient it would be if everyone would ignore any questions of fact in the troubled area of heredity and intelligence, for the sake of peace and harmony, this simply cannot happen. Our society is too committed to the reality principle, at least in areas that bear on our bread-and-butter activities. And the question of the distribution of intellectual potentials arouses intense reactions precisely because it bears on concrete issues in many aspects of social policy. Stent's prescription would only prolong the dispute.[12]

The struggle to reconcile inspiring myths with harsh realities will always be with us. But the political costs of deliberately suppressing objective knowledge, in order to protect a myth, can go far beyond merely prolonging specific disputes. Though such suppression is always based on dedication to what the advocates consider a noble cause, Plato's noble lie all too easily becomes Hitler's big lie—and noble deception all too easily slides over into self-deception.

Is Objective Knowledge Possible?

Having emphasized the value of building on an objective recognition of reality, I shall close by considering briefly recent criticisms of the assumption that we can ever acquire such objective knowledge. One major source of this skepticism is the frustration of social scientists over the difficulty of separating the analytic content of their studies from the frequent policy implications. This topic has been discussed with great good sense by Charles Frankel,[13] and I shall consider only selected aspects.

Many social scientists have thrown up their hands at the problem of eliminating bias, holding that the honest investigator can do no better than to warn the reader by declaring his bias at the outset. Indeed, some even argue that in the social sciences the choice between alternative conclusions should take into account not only evidence and logic but also the anticipated social costs. This view is tempting when we are dealing with conclusions that impinge on our moral convictions. Nevertheless, its acceptance would remove such studies altogether from the realm of science. For the cardinal characteristic of science, and the key to its success, has been the requirement that we abandon any preconception, however treasured, in the face of compelling evidence to the contrary. To a natural scientist the separation of research findings in social science from their application to public policy resembles closely a problem in the natural sciences that we have discussed above: the separation between basic knowledge and control of its applications. In both areas the findings and the logical inferences derived from them can in principle be objective, but policy decisions are then made by a process that also introduces values.

We can sharpen this discussion by considering further the question of the heritability of intelligence, which has become the present test case in the periodic struggle to defend the objectivity of science against political attacks. This question is often treated as one in social science, but it really lies squarely in biology. To be sure, it has important social implications—but so does man's origin, or the distribution of the sickle cell gene. In addition, unlike various monogenic hereditary diseases, any behavioral trait depends very much on the interaction of genes with the environment, and it also depends on so many genes that we cannot locate them individually or identify their molecular products. Accordingly, the available methods can provide only statistical information, rather than identification of individual genotypes. But neither this limitation nor the relevance of the findings for human affairs justifies the assumption that the investigator in behavioral genetics is bound to be influenced by bias, any more than one who studies the distribution of blood group genes: the scientist does not create the reality that he discloses. And the problem of accepting the relevance of behavioral genetics will increase when the advance of neurogenetics eventually permits us to identify individual genetic differences more definitively: Haldane's dream for optimizing education (see note 6).

We should also consider a more far-reaching denial of the objectivity of science, which has been advanced in recent years by certain philosophers and historians of science. This trend seems to have several sources. (1) The possibility of acquiring any objective knowledge has been attacked by neo-Marxists—perhaps to protect ideological dogmas from the danger of contradiction by reality. In an era of disillusion with our social institutions, and of disappointment at the failure of technological progress to solve our social problems, this view seems to have spread widely in academic circles. (2) Our knowledge of reality is acquired by nervous systems that have evolved to deal with a particular range of dimensions; and as Heisenberg has shown in the uncertainty principle, and Einstein in the theory of relativity, phenomena at dimensions outside that range do not fit into the perceptions of time, space, and matter natural for us. By a rather large and unwarranted leap these insights have been thought to weaken confidence that any scientific observations or inferences, even at the level of visible dimensions, correspond reliably to reality. (3) Science as an *activity* involves large subjective elements (and hence value judgments). These elements include the individual's choice of a problem, his choice of what experiments to perform, and society's choice of what work to support. Even more fundamental is the crucial role of hypothesis, for sophisticated students of the philosophy of science (especially Whewell, Peirce, Popper, and Medawar) have recognized that Baconian induction—the spontaneous emergence of a general principle from a collection of observations—fits only the early, descriptive stages of a science. It certainly does not reflect experimental sciences, whose logic involves a continual exchange and feedback between observation, imaginative creation of explanatory hypotheses, logical deduction of additional consequences, and tests of these predictions.

Since science thus has subjective as well as objective elements, it clearly is not value-free in *all* of its aspects. Recognition of this fact, together with preoccupation with the ambiguous term "value-free," has led some to the spurious inference that if values enter anywhere into the activities of a scientist, they also must influence his conclusions: the vaunted objectivity of science is therefore considered a myth. However, this view fails to recognize the dialectic between creation and criticism in science: a methodology that selects, in a sort of Darwinian process of testing and elimination (falsification), the objectively supportable from the false among the many conceptual products of its subjective, creative, value-laden activities. This selection occurs continually in the activities of the individual scientist, and it is further refined through finely honed communal activities.

In addition, we should recognize that scientific conclusions are based ultimately on probabilities, rather than on an absolute causal determinism. Objectivity thus does not imply absolute truth: in principle every conclusion is subject to future refutation or refinement. But in the intellectual edifice constructed by science every addition confirms the reliability (that is, the probability of correspondence with reality) of the foundations on which it builds. I conclude, then, that the biological sciences can indeed reach objective, reliable scientific conclusions about their subject matter. Obviously, at the growing points much that is stated in the name of science does not prove to be verifiable: what science produces might be described in terms of a gradient of reliability.

I am not here concerned with opposing the epistemological analysis of the meaning of objectivity, at the level of technical philosophy: what is disturbing is the denial of objectivity at the level of the interaction of science and society. Such extreme cognitive relativism, like extreme moral relativism, is a dangerous doctrine, easily used to persuade troubled people to replace common sense by nonsensical arguments couched in philosophical terms. The politicization of science thus becomes an avenue to political manipulation in a broader sphere.

Summary and Conclusions

While concern over the possible production of dangerous organisms by recombinant DNA research was in principle legitimate, a reassessment of the controversy suggests that in practice an excessive early involvement of the public, and a lack of confidence in the sense of responsibility of the scientists involved, slowed a reasonable resolution. As an alternative approach I have advocated a two-stage process of evaluating and regulating such potential hazards, with complementary roles for the scientific and general communities.

With respect to concern over knowledge that may give us dangerous powers, I have emphasized that we are not able to foresee the full range of positive and negative consequences of any basic knowledge. We should therefore try to control not the knowledge but those applications whose effects are demonstrably harmful or too costly. In the basic biomedical sciences, in particular, ethical challenges have been invoked for applications that are far too distant for profitable discussion, and in some cases perhaps even impossible: for example, genetic blueprinting of personalities and human cloning.

A third concern, over increased insights into genetic aspects of human nature, is more complex. On the one hand, evidence from this field may conflict painfully with cherished preconceptions. On the other hand, civilization today has a growing sense of crisis—and though this feeling may derive largely from the consequences of the rapid growth of science and technology, we are not likely to improve our responses by refusing to use the power of science. On the contrary, in the long run increased insight into the biological roots of our behavior should aid us in meeting the crisis. If we should cut off the flow of such insights, in order to sidestep immediate problems, we may pay dearly in the long run—not only through deprival of valuable knowledge but also through damage to the ideals of an open society.

All three of the concerns that I have considered involve uncertain conjectures over future possible catastrophes. They therefore lend themselves to demagogic appeal. To truly protect society's interests, and not simply to create the illusion of protection, we must develop better social mechanisms, for assessing and controlling those activities that may be dangerous, and for protecting those that offer promise. But however much we would like to identify incipient dangers early, we cannot expect to see very far ahead. Hence we must continue to rely largely, like the evolutionary process, on trial and error. The real problem is whether we can develop sufficiently rapid corrective mechanisms, in an age when our errors may be so much more costly than in the past.

In a broader perspective, concern over progress in biology is symptomatic of a more general concern about the goals of our society. Even the scientific community has become shaken in its confidence in several former articles of faith: that truth is a supreme end, that progress in science is good for society, and that the scientific community can be trusted with a high degree of autonomy.[14]

The problem is pressing. As Philip Handler recently stated to the National Academy of Sciences: "For better or worse, the terms of a new social contract between the scientific community and the larger society are being forged. It behooves us to help optimize the terms."[15] But this is not easy, when dealing with critics who focus entirely on the dangers or costs of science and technology while taking the benefits for granted.

Perhaps this overreaction will only be part of a historical cycle between romantic interest in good intentions and classic interest in objective truth.[16] The antiscience movement could then prove to be a transitory stage, and even a useful prod, in our struggle to define more clearly what we seek when we speak of progress. For in contrast to our nearly exponential progress in science and technology, progress in a moral sense, or in the sense of overall human welfare, is difficult to demonstrate. Indeed, it is even difficult to define, because it involves normative concepts. We can no longer share the confidence of Condorcet or Spencer or Marx in the perfectibility of man and of society. Though man has increasing control over his fate, the contradictions in his nature, as well as the role of chance, create limits to this control. Tragedy is therefore still inevitable. Perhaps a particularly valuable kind of progress today would be a more realistic recognition of the limits of what we can control, and of the underlying biological reasons for these limits.

The antiscience movement reflects discouragement over the magnitude of our challenges, so amplified by technology: to develop sufficient harmony, between and within nations, to save us from our capacity to destroy civilization; to learn how to use finite resources with foresight; and to broaden the opportunity for fulfillment of individual potentials. Though evolutionary and neurobiological insights will not solve these problems they may help us to manage them better, and to decrease the amount of tragedy in our lives.

Notes and References

1. Gunther S. Stent, *The Coming of the Golden Age* (New York: Doubleday, 1969).
2. M. A. Israel, II, W. Chan, M. A. Martin, and W. P. Rowe, "Molecular Cloning of Polyoma Virus DNA in *Escherichia coli:* Oncogenicity Testing in Hamsters," *Science* 205 (1979): 1140–42.
3. See E. B. Hook, "Geneticophobia and the Implications of Screening for the XYY Genotype in Newborn Infants," in *Genetics and the Law*, ed. A. Milunsky and G. J. Annas (New York: Plenum, 1976) pp. 73–86; and Bernard D. Davis, "XYY: The Dangers of Regulating Research by Adverse Publicity," *Harvard Magazine* 79 (October 1976): 26–30.
4. See T. I. Emerson, "The Constitution and Regulation of Research," in *Regulation of Scientific Inquiry*, ed. K. M. Wulff, AAAS Selected Symposium 37 (Boulder, Colo.: Westview Press, 1979).
5. Nuclear energy might be considered to fall into the class where the peril outweighs the promise. Yet even here, as our civilization is threatened by dwindling energy supplies, only the future will be able to balance the benefits against the risks. And in any case, how could we have arranged for the advance of physics to stop short of this discovery?
6. J. B. S. Haldane, *Heredity and Politics* (New York: Norton, 1938).
7. Ernst Mayr, *Animal Species and Evolution* (Cambridge, Mass.: Harvard University Press, 1963), p. 5.

8. F. A. Hayek (*The Counter-Revolution of Science* [Glencoe, Ill.: Free Press, 1952], p. 15) uses the term "scientism" to denote the error of trying to build the social sciences in "slavish imitation of the method and language of science," rather than simply in its general spirit of disinterested inquiry. But this focus on method leaves out the question of goals and limits.
9. For example, Wilson's striking accomplishment, in launching sociobiology as a major discipline and in forcing a reexamination of the prevalent extreme environmentalism, may be diluted by such concepts as "a biology of ethics" and "a genetically accurate and hence completely fair code of ethics"; E. O. Wilson, *On Human Nature* (Cambridge, Mass.: Harvard University Press, 1978).
10. In the English language the confusion between natural and social sciences is increased by the sharp separation of both from the humanities. In German "*Wissenschaft*" refers to scholarly knowledge, which is divided into the exact, the social, and the humanistic categories.
11. Gunther S. Stent, "The Decadence of Scientism," in his *Paradoxes of Progress* (San Francisco, Calif.: W. H. Freeman, 1978), p. 207. Also in *Foundations of Ethics and Its Relationship to Science*, vol. 2, ed. H. T. Engelhardt and D. Callahan (Hastings, N.Y.: Hastings Institute of Society, Ethics, and the Life Sciences, 1977); and in *The Hastings Center Report*, 6(6) (1976): 32–40.
12. Another essay in Stent's book (*Paradoxes of Progress*, p. 1) illustrates the contradictions created by an effort to defend rationally the deliberate disregard of objective knowledge for the sake of social coherence. In criticizing Peter Medawar's call for continued confidence in progress through science, and advancing his own pessimistic predictions—which are surely less likely to promote present social coherence—his defense is his "accurate assessment of the actual situation."
13. Charles Frankel, "The Autonomy of the Social Sciences," in *Controversies and Decisions: The Social Sciences and Public Policy*, ed. C. Frankel (New York: Russell Sage Foundation, 1976).
14. The changed attitude is reflected in recent actions of several central institutions. (1) In earlier years the U.S. National Academy of Sciences, if faced with the recombinant DNA problem, would doubtless have set up a committee that would have sifted the evidence and produced a presumably authoritative report. In the recent climate, however, the academy felt obligated to set up a forum open to the public. A group of activists, invading the first session, provided more dramatic news for the media than the scheduled speakers. (2) Similarly, in *Science* the news columns on recombinant DNA over a period of two years paid much more attention to the arguments against this research than to the counterarguments, though the latter would seem closer to the parent organization's goal of "the advancement of science." (3) Finally, an editorial in *Nature* (271 [1978]: 391) suggested that we should refrain from acquiring knowledge in certain areas, such as the genetics of intelligence, because it would threaten a just and decent society. However, this journal later did publish a balancing guest editorial (272 [1978]: 390).

15. Philip Handler, Presidential Address, U.S. National Academy of Sciences, April 24, 1979.
16. See Stephen Toulmin, "From Form to Function: Philosophy and History of Science in the 1950s and Now," *Daedalus* (Summer 1977): 143-62.

29

Limits in the Regulation of Scientific Research

The topics of the preceding two papers are taken up again in this one, but with more emphasis on principles that might guide our responses.

According to public opinion polls in the United States, the man in the street still holds science in high regard, however disillusioned he may be about most other social institutions. On the other hand, many social critics insist that we should no longer grant to scientists autonomy in governing their research, because it is increasingly creating serious hazards.

We appear to be dealing with a major shift in attitude, and not a mere fluctation in interest in the problem of the "two cultures." But I believe that most of the criticism is based on a confusion between technology and science, and on a belated recognition of the social and environmental costs of technology. Since science has advanced so effectively, and has made innumerable beneficial social contributions, within a tradition of autonomy, it seems important to examine carefully the reasons for any proposed change in its governance, and the probable consequences. It is also important to analyze separately the different problems presented by different kinds of research.

Loren Graham[1] has presented an elegant analysis of the many distinct categories of concern about science and technology (apart from the irrational ones that will always be waiting in the wings). In areas of

---

Modified from *Ethics for Science Policy*, ed. T. Segerstedt, a Nobel Symposium (New York: Pergamon, 1978).

technology and medicine many criticisms have been justified and useful. They have led us to question the technological imperative and to try to control destructive technology, and to recognize that the medical profession, with benign intent, long ignored the ethical importance of informed consent in research on human subjects. Few medical investigators would question the value of our new sensitivity on this issue. This development has, of course, created new needs: to balance individual rights with society's right to benefit from medical experimentation; and to find means to provide insurance and compensation for those who accept the risks on behalf of society.

In contrast to these real sources of concern, in most areas of fundamental research the alleged hazards to the public are less certain. I would like to focus on three of these areas: research activities that are themselves dangerous because of the materials employed or produced; research whose results might lead to harmful future technological applications; and knowledge that is thought to create a present danger by undermining human values. Of the three, the last may, in the long run, constitute the greatest threat to the integrity of science.

Before proceeding to consider these problems separately, I would note that we may compound our problems if we invoke the concept of ethics too casually. We clearly face real ethical problems in such areas as the rights of human and animal subjects, or the effects of our technology on the environment that we leave for future generations. But I would suggest that the regulation of dangers (or of priorities) in research is in general a pseudo-ethical problem. The distinction is important: for if we stress rights, rather than technical assessment of costs and benefits and political assessment of competing interests, we will encourage the application of unnecessarily absolute principles. The resulting sharper polarization will make it harder to reach a settlement.

Potentially Dangerous Research Activities

In analyzing specific areas, I would like to start by eliminating a false formulation of the problem, which has appeared in some of the more impassioned arguments against the traditional autonomy of science: the claim that scientists are demanding absolute freedom from restraints on their investigative procedures. This claim distorts the concept of "freedom of inquiry." Any such freedom implicitly recognizes the need to set limits on permissible research procedures, whether on grounds of cruelty or human dignity (as noted above for clinical research) or on grounds of safety. For example, there has been no resistance to the requirement that laboratories working with radioactive materials be licensed and subject to regulation and inspection. Moreover, in some kinds of research danger is inescapable: but it is the scientists themselves who are usually most at risk. For example, in the history of laboratory work with pathogenic microbes, over four thousand recorded infections have arisen, with several hundred deaths; and though technical advances since World War II have greatly decreased the risk, a few fatalities from novel viruses are still reported each year. Fortunately for society, there are scientists willing to take such risks.

But dangers voluntarily faced by investigators are not of great concern to the public. What are matters of legitimate concern are hazards that extend potentially to the outside community—for example, research with pathogenic microorganisms that might spread. Yet it is interesting that no cry has arisen for external regulation of laboratories engaged in such work. Two reasons are apparent: we have reliable general knowledge of the hazards; and long experience has generated confidence that professionals will set up reasonable precautions against known dangers. And this confidence has persisted even though we know that human beings are fallible, and that carelessness or chance accidents have initiated a few instances of spread of infection to a small number of contacts outside the laboratory.

It is apparent, then, that what really arouses public anxiety is not hazards *per se*, however serious, but novel, conjectural, uncertain hazards. The uncertainties inevitably lead to divergent views in the scientific community, and if these reach the public they create fear of the unknown. The result is demands for extreme precautions, and for participation of the public in the assessment and regulation of the risk.

What To Do?

Placing scientists on the defensive for achievements that merit pride, and bureaucratizing their activities, will hardly encourage bright young people to enter the field, nor will it promote the effectiveness of those already there. Moreover, many molecular biologists are disillusioned by the consequences of their experiment in public candor. What lessons can we extract from the experience? I cannot offer a program, but I would like to suggest some principles.

(1) Where the facts are not decisive we must base action on *judgment*—and in technically complex areas the judgments of people deeply versed in a field are surely more reliable than those of laymen, however reasonable. Individual scientists, of course, may be biased and even irrational; but a committee of responsible and trained people has so far provided the best available protection against wild judgments.

Contrary to this point of view, some public-interest lawyers and some politicians now assert that the traditional autonomy of scientists in regulating hazards in their research must be halted, because one cannot trust those within a field to regulate themselves. But the universality of this principle is not so evident. In work in pure (as opposed to applied) science, the motives, interest, and rewards of the scientific community, and its tradition of open publication, intense mutual scrutiny and effective peer pressure differ widely from those in the areas—politics, business, and practicing professions—that have generated a justifiable mistrust of self-policing. We need only look at such examples as the control of radioactive materials in laboratories, and the spontaneous concern of the investigators who started the work with recombinant DNA. Indeed, a particularly visible testimonial is provided by the Citizens' Experimentation Review Board of Cambridge, Massachusetts. After nearly one hundred hours of testimony this lay board concluded that the scientists had acted responsibly on their own initiative, and that the public interest would not be threatened by the building of the proposed P3 facility at Harvard.

(2) The scientist's and the layman's points of view also differ in the relative weight attached to the value of *preventing harm* and of *promoting benefits*. As Freeman Dyson[2] has pointed out, discussions of whether or not to allow a particular kind of research express the risk of active harm to individuals in a language congenial to lawyers and familiar to the public, while "the hidden cost of saying no"—deprival of a potentially valuable advance—carries much less weight (except with those deeply immersed in science) and no risk of legal retribution. Part of the difficulty is that the future benefits of a particular piece of research often cannot be specified with assurance; but people with science in their bones know that the sum of the research in a live area, selected and pursued according to the judgment of the scientists, has regularly paid off and can continue to do so.

I therefore suggest that our ethics should place greater weight on positive social obligations. This idea is not novel, though it is hard to convert into legislation. Millenia ago, some societies advanced from the purely negative ethics of sanctions and proscriptions to the positive ethics of "love thy neighbor." The scientific era places on us the burden of developing a parallel ethics and practical politics of research, in which the rights of the individual to protection from harm are balanced with the rights of society to be helped. As the ethicist Joseph Fletcher has stated: "Investigators and subjects, both, owe it to society to learn how to save life, but society owes them, each in his own role, financial and legal support, and protection of their results from abuse by business and yellow journalism. Investigators in particular are morally obligated . . . to accept monitoring by their peers, but not by incompetent 'public watch-dogs' who are not equipped to watch intelligently."[3]

(3) The basic procedural question in decision-making about risks in

research is not *whether* the public should participate, it is *when* and *how*. I would emphasize the need for a two-stage process, with a *complementary* relation between the scientific and the general communities. Technical discussions of possibly risky activities should consider freely all conceivable actions and consequences, including many hypotheses that will eventually be discarded. The presence of the media at this early stage, with their temptation to select sensational items, is bound to impair the integrity of such discussions and the weight subsequently attached to various items. In addition, premature intrusion of the public into the technical phases of the discussion will invite naîve or demagogic confusion of the issues. In the DNA controversy,[4] for example, the presence of the press clearly contributed to an autocatalytic process of maximizing the hypothetical dangers. With a high noise level, who could be heard saying that we were crying wolf? And how could a government official, even if skeptical about the dangers, ignore widespread public anxiety?

Ideally, the first-stage exploration of the problem by responsible professionals would provide a clear and realistic message to the public and its representatives, and they would then participate in the second stage, i.e., in the policy decisions over the acceptable balance of risks and benefits. Such an approach has been spelled out in the greatest detail by Kantrowitz.[5] His proposal has not aroused wide support, perhaps because the unfortunate title "Science Court" seemed to emphasize an adversary process. The heart of the proposal, however, is not this aspect but a separation of the technical and the decision-making processes.

(4) The obverse of scientific responsibility is *public responsibility*. Education of the public about science becomes increasingly urgent: as matters of public policy intersect increasingly with scientific principles ignorance provides fertile ground for sensationalism, and effective participation requires an informed public. Moreover, the experience of the Cambridge Experimentation Review Board demonstrates the ability of laymen to absorb the relevant information and to come to a sensible conclusion. On the other hand, we must recognize that there are serious limits to the interests and the comprehension of the general public in technical discussions. And participatory democracy in such matters is risky: it was only by a narrow margin that the public in Cambridge was represented by an uncommitted board, rather than by the ideologically oriented board preferred by the mayor.

(5) Though we must seek improvement in procedures and in public education, the deepest issue seems to me to be that of restoring *public trust* in the scientific community. That community is accustomed to controversy about the verifiability and the implications of new scientific findings, and it has developed finely honed communal mechanisms for

eventually selecting those additions to the growing edifice that can be built on reliably. The same tradition, applied to questions of hazard, seems to offer society's best safeguard against avoidable catastrophes on the one hand and gratuitous nightmares on the other. Just as a responsible physician would not tell a patient that the diagnostic possibilities in his case ranged from neuritis to cancer, so the scientific community has a responsibility, and should be given discretion, to proceed with care in transmitting conjectural information to an easily alarmed public.

Research That Creates Dangerous Powers

Moving beyond research activities that themselves are dangerous, we can also illustrate with recombinant DNA a second source of anxiety: research whose results give us powers that might be misused in the future. For though most of the overt public reaction to the DNA problem, and the concern of the legislature, centered on the danger of producing harmful organisms, underneath lay another, gnawing fear: that increasing knowledge in this area brings us close to the power to manipulate human genes.

This fear is based on a highly questionable assumption: that if we do reach the goal of gene therapy, governments could then misuse this power to blueprint personalities. In fact, a closer look at the technical aspects of the subject is very reassuring: the gap between gene therapy and genetic control over behavior is enormous.[6,7]

As the technical gap between gene therapy and control of personality continues to make Aldous Huxley's "brave new world" a distant prospect, concern over genetic engineering in man will no doubt dwindle. But future research will give us similar dilemmas about novel capabilities in other areas.

It is therefore important to consider the general problem of research that leads to dangerous powers, and to draw a sharp line between this kind of research and those activities that are themselves dangerous. I have already discussed the latter as a legitimate cause for concern. But when the question is about future powers, I would emphasize the value of regulating at the level of the specific applications, rather than at the level of discovery of the underlying knowledge. My objection is based not on any absolute ethical principle of freedom of inquiry but on the utilitarian consideration that virtually any kind of power can be used for good or for ill: if we really could reliably foresee that the ill would predominate and proscription of particular knowledge would be very sensible; but the history of science furnishes no basis for confidence that we can make that kind of prediction for new insights into nature.

We return, then, to the idea of a complementary relation between the scientific and the general community. If scientists defend the principle of restrictions only on applications and not on new knowledge, and if they are closest to that knowledge, they have a special moral obligation to call attention to possibly harmful applications as soon as they become visible. It is then up to the general society to decide whether and under what conditions a particular technological application may go forward.

But while we recognize the need to try to prevent harmful applications of knowledge as early as possible, we should also recognize that our capacity to do so is limited: often we will be halting recognized harmful practices, rather than preventing foreseeable ones. Evolution, both biological and cultural, proceeds by trial and error. While humans have a power of foresight given to no other animal, this power is still very limited, and we cannot pretend to be able to predict the range of future applications that will flow from a basic discovery. (Think of Faraday and the development of electric power and electronics.) Even the more tangible problem of technology assessment, concerned with the impact of possible but still untried specific applications, has had very limited success.[8] To expect too much from this enterprise is to fall into the trap of "anticipatory democracy"[9]—a prescription for disillusionment.

In a word: mankind has always faced risks, whether in exploring uncharted territories or in trying unfamiliar foods. If our recent success in conquering many malign forces in nature now leads us to seek the security of a world free from novel hazards, and if we forbid exploration of the new kind of unknown territory opened to us by science, we shall not only be condemning ourselves to remain subject to all the present, still unconquered risks; we shall be crushing one of the most admirable expressions of the human spirit.

Disturbing Knowledge About Man

The two problems that I have discussed so far—dangerous actions and dangerous powers—are to a large extent novel products of our era, derived from the recent explosive advance of science and technology. I have treated them as problems largely of technical assessment and pragmatic political action. I would now like to discuss another source of public anxiety, which involves a long-standing, truly ethical issue in new garb: knowledge that is thought to threaten the foundations of public morality.

Lysenko initiated the attack on genetics, on the basis of the then official interpretations of Marxism in the Soviet Union. Today, we see in Western countries a renewed attack, not on all genetics but on human behavioral genetics. And because the abuses of genetics have received enormous attention, while its positive social contributions are virtually unknown, the idea of outlawing much of human behavioral genetics has spread far beyond the small group of scientists who have expressed ideological sympathy for Lysenkoism:[10] it has even appeared in such a central position as an editorial in *Nature*.[11]

Nevertheless, the reality of genetic diversity will be there whether or not we study it scientifically: scientists do not create the truths that they reveal. The problem of whether to face truths about human nature is large, for the potential conflict with our preconceptions will surely grow as our scientific explorations into this area probe deeper—not only in genetics but also in the advancing sciences of neurobiology and sociobiology. If we proscribe such inquiries we will be adopting a particularly parochial view of the nature of justice. For justice is a cultural construction, continually adapting to new material and social circumstances, and certainly capable of adapting to any new insights into human nature. If we arrest our insights in their present state we will, in effect, be subordinating empirical knowledge to normative beliefs, on the false premise that we are obligated to freeze those beliefs.

Thomas Huxley once described Herbert Spencer's definition of a tragedy as a beautiful idea killed by an ugly fact. In a world that has advanced beyond Spencer's Social Darwinism, and has brought science closer to the problems of human nature,[12] we cannot afford to reject the resulting facts as ugly. For if we build policies on assumptions that contradict reality nature will have the last word, and our policies will not be effective or durable. Hence freedom of inquiry in this area, just as in those discussed above, can be defended on long-term pragmatic grounds—as well as on grounds of admiration for the cultural value of creative intellectual activity. But at the same time, while we are defending the value of science and of freedom of inquiry, we must also avoid the scientism that expects science to solve our social problems. It cannot do so, because it cannot answer questions of values. What science can do is help us to build on reality, by giving us access to facts that intersect with ou values and that limit our options.

Summary and Conclusions

It is commonly asserted today that many areas of scientific research pose threats to the public interest and must be restrained by the public. I have argued that this conception of science is false, and that external regulation is required for technological applications but not for scientific research.

Specifically, I have rejected the idea that we should forbid certain kinds of knowledge because they will lead to harmful applications. One reason is the overwhelming historical evidence that any knowledge can have both good and bad applications, and that we cannot see their distribution far in advance. Another reason is the conviction that man's curiosity and his drive to develop better tools are too deep-seated to be permanently suppressed.

I have also rejected the idea of forbidding kinds of knowledge (e.g., about human genetics) that are thought to threaten a decent society. The threat is largely inferred from the disastrous consequences of early political extrapolations of evolutionary and genetic theory. But though this history still causes understandable apprehension, the modern scientific advances of sociobiology and behavioral genetics find a much more apt model in population genetics, which played a most valuable, though unrecognized, social role by eliminating earlier destructive misconceptions about the nature of race. In addition, blocking off knowledge about the reality of our biological legacy will deprive us of the benefits of that knowledge and will not protect us from its feared implications: the reality will be there, and it will influence the success or failure of our policies, whether we recognize it or not.

A third group of scientific activities is a more legitimate source of concern: those that themselves may be dangerous. Among these, the real ethical problems of research on human subjects are now well in hand; and with laboratory materials that present clearly defined dangers there has been no problem. But with unknown, conjectural risks we have a serious problem. Scientists will need time to thrash out the arguments, and to collect the new information required for reaching a reasonable consensus. Meanwhile, as the recombinant DNA episode illustrates, a public prematurely exposed to these technical discussions will inevitably be confused and alarmed. The results may include a large waste of resources, the growth of an onerous bureaucracy, and regulations that hamper scientific advance in the effort to protect against nonexistent hazards.

Increased scientific education of the public will clearly improve the decision-making process. Scientists have a special responsibility here—particularly in educating the public about the nature of science and the conditions that promote its productivity. But even with an ideally educated public it would still be desirable to separate, as far as possible, the stage of technical assessment by the scientific community from the stage of ultimate decisions by the public and its representatives.

But while better procedures should help us to avoid a repetition of the DNA hysteria, I believe we will also need a restoration of trust in the scientific community. For scientists have had a remarkable record of professional honesty and responsibility—not because they are inherently more virtuous than other citizens, but because neither they nor the world gain from their research unless their actions and findings are made public, subjected to minute scrutiny by their peers, and found to be verifiable. Moreover, as an intensely communal, critical activity, science has developed finely honed mechanisms for evaluating controversies dispassionately, and with emphasis on evidence and logic rather than on rhetorical skill and public stature. To be sure, the collective judgment of the scientific community, or of its traditional committees of trusted experts, is not infallible: but it would be difficult to find a social institution with a better record of success in winnowing truth from falsehood.

Of late, perhaps because of guilt over the charge of élitism and the ills of technology, some scientists appear to be losing confidence in the objectivity of scientific knowledge and in the ability or the right of their community to speak with any authority. But while there is no room for absolute authority in science, there is also no room for extreme intellectual relativism. In the areas of its expertise the scientific community has the authority, and the obligation, to help the public to discriminate between rational and irrational views.

NOTES AND REFERENCES

1. L. R. Graham, "Concerns about Science and Attempts to Regulate Inquiry," *Daedalus* (Spring 1978):1.
2. F. Dyson, "On the Hidden Cost of Saying No," *Bulletin of the Atomic Scientists* 31 (1975):23.
3. J. Fletcher, "Ethical Considerations in Biomedical Research Involving Human Beings," in *Proceedings of the International Conference on the Role of the Individual and the Community in the Research, Development, and Use of Biologicals* (Geneva, Switzerland: World Health Organization, 1976).
4. J. D. Watson, "In Defense of DNA," *New Republic* (June 25, 1977):11.
5. A. Kantrowitz, "Controlling Technology Democratically," *American Scientist* 63 (1975):505.
6. J. Lederberg, "Biological Innovation and Genetic Intervention," in J. A. Behnke, ed., *Challenging Biological Problems* (Oxford, England: Oxford University Press, 1972):7.
7. B. D. Davis, "Prospects for Genetic Intervention in Man," *Science* 170 (1970):1279.
8. E. W. Lawless, *Technology and Social Shock* (New Brunswick, N.J.: Rutgers University Press, 1977).
9. S. Weiner and A. Wildavsky, "The Prophylactic Presidency," *The Public Interest* (Summer 1978):3.
10. R. C. Lewontin and R. Levins, "The Problem of Lysenkoism," in H. Rose and S. Rose, eds., *The Radicalization of Science* (London: Macmillan, 1976).
11. Editorial, "Truth at Any Price," *Nature* 271 (1978):391.
12. E. O. Wilson, *On Human Nature* (Cambridge, Mass.: Harvard University Press, 1978).

30

Medical Research and the Law

A symposium on genetics and the law afforded me an opportunity to discuss recent legal restrictions on valuable medical research.

Some degree of public control over science has been both traditional and inevitable. If society pays the bill it has to decide among competing demands for its funds. But to a very large extent the public has been willing to leave decisions about the relative value of various projects, and about their moral justifiability, in the hands of the scientific community, since technical considerations usually loom large in these decisions. Moreover, the public has accepted the view that the interest of scientists in advancing knowledge generally coincides with the interests of society, and that the biases and self-interest of individual scientists are best detected and corrected by their peers. Hence legislators, guided by scientist administrators, have generally determined only how much money to appropriate and have left the apportionment and the mechanisms of distribution entirely up to peer groups.

The last few years, however, have seen an increasing demand for public input into these decisions. The abruptness of this development, and the accompanying criticisms of the scientific community, have come as a surprise. Yet some such reaction can be now seen as inevitable, as science has become more visible and expensive, as it continues to generate novel ethical problems, and as we belatedly recognize that the technological applications of science yield not only benefits but also environ-

In *Genetics and the Law*, ed. A. Milunsky and G. J. Annas (New York: Plenum, 1975.)

mental and social costs. The reaction was intensified, and perhaps precipitated, by several other social developments that coincided with it: the recognition that technological advances present costs and dangers as well as benefits, widespread mistrust of social institutions, heightened concern with the rights of individuals, and antagonism to a vaguely defined elitism. This last development inevitably extended to science, since that enterprise must emphasize respect for intellectual excellence and for objective standards.

The reaction of the scientific community has been one of dismay. It is natural that a group accustomed to self-regulation will resist and resent efforts to introduce external sources of restraint. This is particularly true of the biomedical sciences, proud of their humanitarian goals and accustomed to trust in the judgment of physicians. Yet we must recognize that the change in the public attitude is no passing fad, and we cannot expect to return to the earlier pattern. Moreover, few would deny that the new pressures have had some tangible benefits. For even though the record would no doubt show that most medical investigators have been responsible and conscientious in trying to balance the interests of the subjects of research and those of society, they have acted largely on the unexamined traditions of an earlier, more stratified society. Hence recognition of the need for informed consent where real risk is involved, and recognition of the need for research review committees in hospitals, represent real gains.

Nevertheless, I have deep apprehension, like many of my colleagues, at the prospect of overreactions, and a loss of perspective, that will have serious consequences. This is not a distant specter. As current examples we have the recent ban on fetal research and even limitations on the study of waste materials from a patient. In a particularly ironical twist, a cell biologist at Massachusetts Institute of Technology was using the by-products of circumcisions as a convenient source of cells from a young individual, and because the surgeon did not wish to get tangled up with the problem of obtaining adequately informed consent the investigator found his supply cut off.

The danger of foolish restrictions on research seems especially great in the medical area, for several reasons. First, in the field of medical ethics it is particularly easy to arouse an emotional reaction by presenting a one-sided, simplistic analysis of a complex problem: the XYY story is a recent example. Second, the widespread dissatisfaction of the public with the quality, availability, and economics of medical care is easily projected upon the highly visible teaching and research centers. Third, the problems of biomedical research have attracted the attention of public-interest lawyers, who bring with them the familiar adversary process. Hence instead of the traditions of scientific discussion, in which each participant is expected to try to consider all the evidence, we find unpleasant, polarized exchanges between prosecutors and defendants. This atmosphere was conspicuous in a recent National Academy of Sciences Forum on the ethics of medical research, and the results did not seem very constructive.

The ancient adversary tradition of the law may conceivably be the best way to resolve questions of guilt, where the crime is well defined, but it is hardly the best way for society to decide what is licit or illicit in medical research. The greatest danger is that the attacks by public-interest groups will lead not simply to closer regulation by governmental agencies concerned with science, but to direct intervention by legislators, who are quick to seize on issues that catch the public fancy. In areas involving complex technical considerations, not well understood by the public, the result is likely to be premature, heavy-handed legislation, such as the laws passed in many states a few years ago requiring screening for the sickle-cell gene. In this case the laws did not persist very long, for they resulted in direct public actions whose harmful consequences were easily recognized. However, equally foolish legislation restricting scientific research might last much longer, for sins of omission are much less visible than sins of commission.

31

The Adversary Approach to Biomedical Research

Here I focus on the difficulties and the importance of adapting our legal system, with its traditional emphasis on proper form and on the adversary process, to the more pragmatic approach that science cultivates. Other problems concerning the interaction of the law and science will be discussed in Part Six of this volume.

I would like to review briefly the ethical problems and the dangers of overreaction in two areas: fetal research, and experimentation on human subjects in general.

Vigorous objection to fetal research in recent years has resulted in a national moratorium that is preventing most such research today, except that involving tissues from an already dead fetus. In Massachusetts, recent legal actions are causing hospitals and doctors to be even more careful than the law requires, because nobody knows what interpretations are going to be attached to any action involving a fetus. The issue is highly charged because it is so closely linked to the problem of abortion and, hence, to very strong moral convictions on the part of many people.

There is widespread agreement that research on tissues from an already dead fetus is no different from research that may be permitted on adult cadavers. What will be permitted beyond this in Massachusetts is not clear. There has been extensive discussion between supporters of

In *Ethics of Teaching and Scientific Research*, ed. S. Hook, P. Kurtz, and M. Todorovich (Buffalo, N.Y.: Prometheus Books, 1977).

fetal research and some legislators, and the results seem to have been educational. I mention this because it offers hope that sufficient communication can get people with diametrically opposed positions on such issues to see each other's point of view, and to adopt a better compromise. It looks as though the legislation will permit research on a fetus at *the time of* a legal abortion. For example, an investigator might want to test a diagnostic procedure that could lead to useful recognition of fetuses with serious hereditary disease. If the procedure involved is first tested on individuals about to be aborted it can be perfected without exposing wanted infants to the high risk of a novel procedure.

There is stronger objection to another kind of fetal research: one in which procedures are carried out, or drugs are given to a mother, a few hours or a few days ahead of a scheduled abortion. Even though it is recognized that somebody has to take the risk if you are going to test any new drug or any new procedure, and it seems logical to have the risk applied preferentially to fetuses that are scheduled for destruction, the groups opposing this want to preserve for the mother the right to change her mind up to the last moment. Any procedure that might irrevocably commit the mother to an already planned abortion is probably going to be legally blocked. In an excellent article, Gaylin and Lappe, in the *Atlantic Monthly* (May 1975), emphasize that if we do not allow unwanted fetuses to be used we are going to condemn a great many wanted fetuses to undergo the risk.

The second, broader problem is that of human subjects in general in relation to medical research. Those of us in the medical profession have been astounded by the violence of the recent objection in large segments of society to many aspects of medical research. Indeed, biomedical ethics has become a major growth industry. This development is related to the consumer movement, with its emphasis on individual rights and protection from institutions; it reflects loss of confidence in institutions in general; and it reflects the fact that the tremendously increased scale of medical research over the last two decades, with funds appropriated by the government, has involved many more people. Indeed, with this large scale there are inevitably incidents in which things are done that should not have been done. (In addition, however, an atmosphere has arisen in which many investigators might be described as more enthusiastic than sensitive. It is, therefore, a good thing that we are being pressed to be more aware of this problem and to set up committees in hospitals and medical schools to review research projects for their adherence to ethical guidelines.) Finally, a great deal of the reaction arises from a small number of horror stories that have received a great deal of attention in the press. Among them are the Tuskegee study, in which treatment was withheld from some black patients with syphilis, and injections of cancer cells in moribund patients.

It seems quite clear that we cannot solve the problem by saying that only those of us with a medical background can really understand the complexity of the problem and so we should be allowed, now that we have been shaken up, to police ourselves again. It seems that there is too much concern on the part of too many other people. But I do think that the directions in which we are moving also present grave dangers. Initially some philosophers or theologians, with a serious and technical background in medical ethics, began a systematic inquiry into this field, but the discussion has become increasingly strident.

The best statement I have seen on this subject is one by Dr. Franz Ingelfinger, editor-in-chief of the *New England Journal of Medicine*, who has recently written a paper called "The Unethical in Medical Ethics" (*Annals of Internal Medicine* 83 [1975]: 264). He points out that along with the value of having our attention called to the need for more awareness of the problem, there are also very serious dangers, which he describes under five headings. First is the righteousness with which people in the current climate of ethical opinion point the finger at horrors of the past. This action is no more justified than it would be to accuse a current Southern landowner of being responsible for previous Southern landowners who thought it ethical to have slaves. Ethical principles change, and there is a great deal of righteousness in the horror stories that are getting so much attention about medical ethics and medical research. Second, this reaction is amplified by the desire of media for sensationalism. Third, a more basic problem is how to balance concern over individual rights with concern over society's needs. In the present climate of opinion there are large groups who feel that there should be no more research on prison inmates because they cannot ever be considered to have given a truly voluntary consent. I believe this is an overreaction to a situation in which indeed there have been abuses. These could be better corrected by having committees of disinterested people, separate from the researchers, supervising the nature of the consent.

The fourth and fifth dangers that Dr. Ingelfinger mentions are those of bureaucratization and trivialization. If we have enormously detailed rules, and a dilution of real issues with innumerable trivial issues, the whole field becomes very unattractive. If we continue in this direction highly motivated people may be driven away from the field, since more energy, time, and money will go into filling out forms than into actions that are of benefit to the patient and to society.

There is a deep irony in all of this. For while researchers in teaching and research institutions have occasionally caused harm by actions that should be prevented, the general level of medical care in those institutions is undoubtedly far better than that in general practice in this country. The high visibility of these institutions could divert attention from the problems of availability and quality of medical care in general.

One product of recent criticism of research practices has been insistence on informed and voluntary consent. While the principle is excellent, developing effective procedures is not easy. What is information to a doctor in a consent form may also mean a lot to a highly educated patient but may mean little to another patient. Some people who are very concerned with this problem seem to me to be projecting onto the whole class of patients practices they would like for themselves, but that really do not fit most patients. Moreover, the problem of informed consent also gets pushed to an extreme when it is required for procedures that do not involve risk or serious invasion of privacy. Today an investigator must have written informed consent in order to use waste products of a patient that are collected in the hospital.

Perhaps the most general problem in medical research is that of risk-benefit analysis. In any new procedure there is, by definition, some risk involved. Who is going to take on that risk? One of the most valuable things that public-interest lawyers can do in this area, I think, is to try to provide mechanisms for society to insure people for indemnity if they are harmed as a result of accepting this risk. But an added problem of social justice arises if one social class provides the guinea pigs for the benefit of another. That traditionally has been the way medical research was largely done, on the assumption that the risk of harm from research was traded for the much larger risk from less skilled medical care outside the teaching institutions. This is one area in which things clearly have to change. Unfortunately, at present the discussion seems unsatisfactory because much of it is contributed by a group of public-interest lawyers, acting as prosecutors against a medical research community whom they place in the dock. At a forum on medical experimentation at the National Academy of Sciences recently the lawyers and the medical investigators talked completely past each other. One reason for the lack of real communication was the adversary process, which is the natural approach of lawyers but is foreign to physicians and scientists. To many of us this approach seems to inject into the situation an atmosphere unconducive to the kinds of solution we are accustomed to seek—solutions with more emphasis on practicality, and less on form.

Part Six

Genetic Engineering

32

Prospects for Genetic Intervention in Man

In conjunction with the 1969 meeting of the American Association for the Advancement of Science the Department of Biology and the School of Theology at Boston University organized one of the earliest conferences on what was to become a major area of public concern: the social impact of future advances in biomedical sciences. Sir Peter Medawar was scheduled to speak on the prospects for gene manipulation, but owing to illness he was unable to attend. I filled in (though rather diffidently), thus becoming initiated as a writer on issues of science and society.

Just before the conference the issue gained a good deal of public attention. Two colleagues in my department at Harvard Medical School, Jonathan Beckwith and James Shapiro, held a press conference in which they expressed regret at their success in being the first to isolate a gene, in bacteria, because this was a step toward manipulating genes in man. They argued that the prospect of such powers increased the need for radical change in our political structure. However, the news media were much less interested in their political opinions than in a man-bites-dog story: "Scientists say that their science is dangerous." The protagonists learned quickly what got attention. Subsequently, a larger group of scientific activists was formed, Science for the People; building on this experience, they concentrated their efforts on attacking various aspects of genetics, rather than on broader political and economic issues.

My assignment was to assess the technical possibilities, and the likely benefits and dangers, for each kind of genetic intervention. While most of the views expressed in this paper still seem sound to me, the crystal ball, as usual, had some cloudy areas. I could not have foreseen the development of the fantastic recombinant DNA methodology, and so I did not imagine that gene therapy in humans would come so close in as little as fifteen

Science 170 (1970):1279–1283.

years. I also erred in the other direction: it seemed likely then that cloning, already achieved in tadpoles, would soon be extended to mammals; but with subsequent scientific developments that prospect seems increasingly unlikely.

Later this paper had an unexpected consequence. In the discussion of cloning (which my assignment required), I suggested that if this technique should become feasible in other mammals it would be "tempting" to extend it to man, "on the grounds that enrichment for proved talent by this means might enormously enhance our culture." I then coupled this statement with a long paragraph spelling out a number of serious objections. Subsequently, two sensational books grossly distorted the passage.

In one of these, In His Image *by David Rorvik, the citation, which was also used in an advertisement, replaced "would be tempting" by the statement that I "favored" extension to humans. Even more, the citation omitted all of my subsequent caveats. In the face of such a distorted interpretation I felt obligated to look into the possibility of legal action, especially since the book had been produced by a major publisher. However, the lawyer whom I consulted was a gentle soul who persuaded me that a suit would not accomplish enough to be worth all the trouble. Accordingly, even though it seemed worthwhile to try to discourage reputable publishers from accepting books that were patently fraudulent and that distorted the image of science, I found the legal system discouraging, and I settled for the publisher's agreement to remove the offending material from future printings. A successful suit was later brought by a British scientist, on the weaker grounds of invasion of privacy.*

My negotiations with the publisher revealed that Rorvik had never seen my Science *paper but had acquired the mutilated citation from an earlier book,* Who Should Play God?, *by Ted Howard and Jeremy Rifkin. I had previously ignored this title as an obvious piece of alarmist propaganda, but I now realized that it contained the citation in the form copied by Rorvik, and I found that it had sold about 400,000 copies. This widely distributed distortion of my position seemed even more to justify proceedings against the authors. Again, however, I chose a gentle lawyer, who was more interested in civil liberties than in the possible damage done to my reputation, and when he failed to pursue the case vigorously I dropped it.*

Unfortunately, I could not foresee that Rifkin would later succeed in using the courts to impede valuable research in biotechnology. I sometimes wonder whether a successful suit might not have had some restraining influence on his subsequent activities. In any case, Rifkin, as well as Science for the People, *has provided the stimulus for a number of the pieces in this volume.*

Extrapolating from the spectacular success of molecular genetics, a number of essays and symposia[1] have considered the feasibility of various forms of genetic intervention[2] in man. Some of these statements, and many articles in the popular press, have tended toward exuberant,

Promethean predictions of unlimited control and have led the public to expect the blueprinting of human personalities. Most geneticists, however, have had more restrained second thoughts.

Nevertheless, recent alarms about this problem have caused wide public concern, and understandably so. With nuclear energy threatening global catastrophe, and with so many other technological advances visibly damaging the quality of life, who would wish to have scientists tampering with man's inner nature? Indeed, fear of such manipulation may arouse even more anxiety than fear of death. The mass media have accordingly welcomed sensational pronouncements about the dangers.

While such dangers clearly exist, it also seems clear that some scientists have dramatized them[3] in an effort to persuade the public of the need for radical changes in our form of government.[4] But however laudable the desire to improve our social structure, and however urgent the need to guard against harmful uses of science and technology, exaggeration of the dangers from genetics will inevitably contribute to an already distorted public view. Indeed, irresponsible hyperbole on the genetic engineering issue has already influenced the funding of research.[5] It therefore seems important to try to assess objectively the prospects for modifying the pattern of genes of a human being. But let us first note two genetic principles that must be taken into account.

Relevant Genetic Principles

Polygenic Traits and Behavioral Genetics

The recognition of a gene, in classical genetics, depends on following the distribution of two alternative forms (alleles) from parents to progeny. In the early years of genetics, after the rediscovery of Mendel's laws in 1900, this analysis was possible only for those genes that exerted an all-or-none control over a corresponding monogenic trait—for example, flower color, eye color, or a hereditary disease such as hemophilia. The study of such genes has continued to dominate genetics. However, monogenic traits constitute only a small, special class. Most traits are polygenic: that is, they depend on multiple genes. Moreover, each gene itself is polymorphic—that is, it is capable of existing, as a result of mutation, in a variety of different forms (alleles); and though the protein products of these alleles differ only slightly in structure, they often differ markedly in activity. Accordingly, these traits vary continuously rather than in an all-or-none manner.

For our purpose it is especially pertinent that the most interesting human traits—relating to intelligence, temperament, and physical structure—are highly polygenic. Indeed, man undoubtedly has hundreds of thousands of genes for polygenic traits, compared with a few hundred recognizable through their control over monogenic traits. However, the study of polygenic inheritance is still primitive; and the difference from monogenic inheritance has received little public attention. Education on the distinction between monogenic and polygenic inheritance is clearly important if the public is to distinguish between realistic and wild projections for future developments in genetic intervention in man.

Interaction of Heredity and Environment

The study of polygenic inheritance is difficult in part because it requires statistical analysis of the consequences of reassortment, among the progeny, of many interacting genes. In addition, even a full set of relevant genes does not fixedly determine the corresponding trait. Rather, most genes contribute to determining a *range of potential* for a given trait in an individual, while his past and present environments determine his phenotype (that is, his actual state) within that range. At a molecular level the explanation is now clear: the structure of a gene determines the structure of a corresponding protein, while the interaction of the gene with subtle regulatory mechanisms, which respond to stimuli from the environment, determines the amount of the protein made. Hence, the ancient formulation of the question of heredity versus environment (nature versus nurture) in qualitative terms has presented a false dichotomy and has led only to sterile arguments.

Possibilities in Genetic Manipulation

Somatic Cell Alteration

Bacterial genes can already be isolated[6] and synthesized;[7] and while the isolation of human genes still appears to be a formidable task, it may also be accomplished quite soon. We would then be able to synthesize and to modify human genes in the test tube. However, the incorporation of externally supplied genes into human cells is another matter. For while small blocks of genes can be introduced in bacteria, either as naked DNA (transformation) or as part of a nonlethal virus (transduction), we have no basis for estimating how hard it will be to overcome the obstacles to applying these methods to human cells. And if it does become possible to incorporate a desired gene into some cells, in the intact body, incorporation into all the cells that could profit thereby may well remain difficult. It thus seems possible that diseases depending on deficiency of an extracellular product, such as insulin, may be curable long before the bulk of hereditary diseases, where an externally supplied gene can benefit only those defective cells that have incorporated it and can then make the missing cell component.

Such a one-shot cure of a hereditary disease, if possible, would clearly be a major improvement over the current practice of continually supplying a missing gene product, such as insulin. (It could be argued that improving the soma in this way, without altering the germ cells, would help perpetuate hereditary defectives; but so does conventional medical therapy.) The danger of undesired effects, of course, would have to be evaluated, and the day-to-day medical use of such material would have to be regulated: but these problems do not seem to differ significantly from those encountered with any novel therapeutic agent.

Germ Cell Alteration

Germ cells may prove more amenable than somatic cells to the introduction of DNA, since they could be exposed in the test tube and therefore in a more uniform and controllable manner. Another conceivable approach might be that of *directed mutagenesis:* the use of agents that would bring about a specific desired alteration in the DNA, such as reversal of a mutation that had made a gene defective. So far, however, efforts to find such directive agents have not been successful: all known mutagenic agents cause virtually random mutations, of which the vast majority are harmful rather than helpful. Indeed, before a mutagen could be directed to a particular site it would probably have to be attached first to a molecule that could selectively recognize a particular stretch of DNA;[8] hence a highly selective mutagen would have to be at least as complex as the material required for selective genetic recombination.

If predictable genetic alteration of germ cells should become possible it would be even more useful than somatic cure of monogenic diseases, for it could allow an individual with a defective gene to generate his own progeny without condemning them to inherit that gene. Moreover, there would be a long-term evolutionary advantage, since not only the immediate product of the correction but also subsequent generations would be free of the disease.

Genetic Modification of Behavior

In contrast to the cure of specific monogenic diseases, improvement of the highly polygenic behavioral traits would almost certainly require the replacement, in germ cells, of a large but specific complement of DNA. Since I find such replacement, in a controlled manner, very hard to imagine, I suspect that such modifications will remain indefinitely in the realm of science fiction, like the currently popular extrapolation from the transplantation of a kidney or a heart, with a few tubular connections, to that of a brain, with hundreds of thousands of specific neural connections.

Copying by Asexual Reproduction (Cloning)

We now know that all the differentiated somatic cells of an animal (those from muscle, skin, and the like) contain, in their nuclei, the same complete set of genes. Every somatic cell thus contains all the genetic information required for copying the whole organism [see comment preceding this article for correction]. In different cells different subsets of genes are active, while the remainder are inactive. Accordingly, if it should become possible to reverse the regulatory mechanism responsible for this differentiation any cell could be used to start an embryo. The individual could then be developed in the uterus of a foster mother, or eventually in a glorified test tube, and would be an exact genetic copy of its single parent. Such asexual reproduction could thus be used to produce individuals of strictly predictable genetic endowment; and there would be no theoretical limit to the size of the resulting clone (that is, the set of identical individuals derivable from a single parent and from successive generations of copies).

Though differentiation is completely reversible in the cells of plants (as in the transfer of cuttings), it is ordinarily quite irreversible in the cells of higher animals. This stability, however, depends on the interaction of the nucleus with the surrounding cytoplasm; and it is now possible to transfer a nucleus, by microsurgery or cell fusion, into the cytoplasm of a different kind of cell. Indeed, in frogs differentiation has been completely reversed in this way: when the nucleus of an egg cell is replaced by a nucleus from an intestinal cell embryonic development of the hybrid cell can produce a genetic replica of the donor of the nucleus.[9] This result will probably also be accomplished, and perhaps quite soon, with cells from mammals. Indeed, there is considerable economic incentive to achieve this goal, since the copying of champion livestock could substantially increase food production.

Another type of cloning can already be accomplished in mammals: when the relatively undifferentiated cells of an early mouse embryo are gently separated each can be used to start a new embryo.[10] A large set of identical twins can thus be produced. However, they would be copies of an embryo of undetermined genetic structure, rather than of an already known adult. This procedure therefore does not seem tempting in man, unless the production of identical twins (or of greater multiplets) should develop special social values, such as those suggested by Aldous Huxley in *Brave New World*.

Predetermination of Sex

Though no one has yet succeeded in directly controlling sex by separating XX and XY sperm cells, this technical problem should be soluble. Moreover, in principle it is already possible to achieve the same objective indirectly by aborting embryos of the undesired sex: for the sex of the embryo can be diagnosed by tapping the amniotic fluid (amniocentesis) and examining the cells released into that fluid by the embryo.

Wide use of either method might cause a marked imbalance in the sex ratio in the population, which could lead to changes in our present family structure (and might even be welcomed in a world suffering from overpopulation). Alternatively, new social or legal pressures might be developed to avert a threatened imbalance.[11] But though there would obviously be novel social problems, I do not think they would strain our powers of social adaptation nearly as much as some urgent present problems.

Selective Reproduction

A discussion of the prospects for molecular and cellular intervention in human heredity would be incomplete without noting that any society wishing to direct the evolution of its gene pool already has available an alternative approach: selective breeding. This application of classical, transmission genetics has been used empirically since Neolithic times, not only in animal husbandry, but also, in various ways (for example, polygamy, *droit de seigneur*, caste system), in certain human cultures. Declaring a moratorium on genetic research, in order to forestall possible future control of our gene pool, would therefore be locking the barn after the horse was stolen.

Having reviewed various technical possibilities, I would now like to comment on the dangers that might be presented by their fulfillment and to compare these with the consequences of efforts to prevent this development.

Evaluation of the Dangers

Gene Transfer

I have presented the view that if we eventually develop the ability to incorporate genes into human germ cells, and thus to repair monogenic defects, we would still be far from specifying highly polygenic behavioral traits. And with somatic cells such an influence seems altogether excluded. For though genes undoubtedly direct in considerable detail the pattern of development of the brain, with its network of connections of 10 billion or more nerve cells, the introduction of new DNA following this development clearly could not redirect the already formed network; neither could we expect it to modify the effect of learning on brain function.

To be sure, since we as yet have little firm knowledge of behavioral genetics we cannot exclude the possibility that a few key genes might play an especially large role in determining various intellectual or artistic potentials or emotional patterns. But even if it should turn out to be technically possible to tailor the psyche significantly by the exchange of a small number of genes in germ cells, it seems extremely improbable that this procedure would be put to practical use. For it will always be much easier, as Lederberg[12] has emphasized, to obtain almost any desired genetic pattern by copying from the enormous store already displayed in nature's catalog.

While the improvement of cerebral function by polygenic transfer thus seems extremely unlikely, one cannot so readily exclude the technical possibility of impairing this function by transfer of a monogenic defect. And having seen genocide in Germany and massive defoliation in Vietnam, we can hardly assume that a high level of civilization provides a guarantee against such an evil use of science. However, several considerations argue against the likelihood that such a future technical possibility would be converted into reality. The most important is that monogenic diseases, involving hormonal imbalance or enzymatic deficiencies, produce gross behavioral defects, whose usefulness to a tyrant is hard to imagine. Moreover, even if gene transfer is achieved in cooperating individuals, an enormous social effort would still be required to extend it, for political or military purposes, to mass populations. Finally, in contrast to the development of nuclear energy, which arose as an extension of already accepted military practices, the potential medical value of gene transfer is much more evident than its military value; hence a "genetic bomb" could hardly be sprung on the public as a secret weapon. Accordingly, we are under no moral obligation to sacrifice genetic advances now in order to forestall such remote dangers: if and when gene transfer in man becomes a reality there would still be time to assert the cultural and medical traditions that would promote its beneficial use and oppose its abuse.

This last obstacle would be eliminated if it should prove possible to develop a virus that could be used to infect a population secretly with specific genes, and it is the prospect of this ultimate horror that seems to cause most concern. However, for reasons that I have presented above the technical possibility of usefully modifying personality by infecting germ cells seems extremely remote, and the possibility of doing so by infecting somatic cells in an already developed individual seems altogether excluded. These fears thus do not seem realistic enough to help guide present policy. Nevertheless, the problem cannot be entirely ignored: in a country that has recently been embarrassed by its accumulation of rockets containing nerve gas even the remote possibility of handing viral toys to Dr. Strangelove will require vigilance.

Genetic Copies

If the cloning of mammals becomes technically feasible its extension to man will undoubtedly be very tempting, on the grounds that enrichment for proved talent by this means might enormously enhance our culture, while the risk of harm seemed small. Since society may be faced with the need to make decisions in this area quite soon, I would like to offer a few comments in the hope of encouraging public discussion.

On the one hand, in fields such as mathematics or music, where major achievements are restricted to a few especially gifted people, an increase in their number might be enormously beneficial—either as a continuous supply from one generation to another or as an expanded supply within a generation. On the other hand, a succession of identical geniuses might exert an excessively conservative influence, depriving society of the richness that comes from our inexhaustible supply of new combinations of genes. Or genius might fail to flower, if its drive depended heavily on parental influence or on cultural climate. And in the literary, social, and political areas the cultural climate surely plays so large a role that there may be little basis for expecting outstanding achievement to be continued by a scion. The world might thus be quite disappointed by the contributions of another Tolstoy, Churchill, or Martin Luther King, or even another Newton or Mozart. Moreover, though experience with monozygotic twins is somewhat reassuring, persons produced by copying might suffer from a novel kind of "identity crisis."

Though our system of values clearly places us under moral obligation to do everything possible to cure disease, there is no comparable basis for using cloning to advance culture. The responsibility for initiating such a radical departure in human reproduction would be grave, and surely many will feel that we should not do so. But I suspect that it would be impossible to enforce any such prohibition completely: the potential gain seems too large, and the procedure would require the cooperation of only a very small group of people. Hence whatever the initial social consensus, I suspect that a stable attitude would not emerge until after some early tests, whether legal or illegal, had demonstrated the magnitude of the problems and of the gains.

A much greater threat, I believe, would be the use of cloning for the large-scale amplification of a few selected individuals. Who would wish to send a child to a school with a large set of identical twins as his classmates? Moreover, the success of a species depends not only on its adaptation to its present environment but also on its possession of sufficient genetic variety to include some individuals who could survive in any future environment. Hence if cloning were extended to the point of markedly homogenizing the population, it could create an evolutionary danger. However, we have already lived for a long time with a similar possibility: any male can provide a virtually limitless supply of germ cells, which can be used in artificial insemination; yet genetic homogenization by this means has not become the slightest threat. Since cloning is unlikely to become nearly so easy it is difficult to see a rational basis for the fear that its technical possibility would increase the threat.

Implications for Genetic Research

Though the dangers from genetics seem to me very small compared with the immense potential benefits, they do exist: its applications could conceivably be used unwisely and even malevolently. But such potential abuses cannot be prevented by curtailing genetic research. For one thing, we already have on hand a powerful tool (selective breeding) that could be used to influence the human gene pool, and this technique could be used as wisely or unwisely as any future additional techniques. Moreover, since the greatest fear is that some tyrant might use genetic tools to regulate behavior, and especially to depress human potential, it is important to note that we already have on hand pharmacological, surgical, nutritional, and psychological methods that could generate parallel problems much sooner. Clearly, we shall have to struggle, in a crowded and unsettled world, to prevent such a horrifying misuse of science and to preserve and promote the ideal of universal human dignity. If we succeed in developing suitable controls we can expect to apply them to any later developments in genetics. If we fail—as we may—limitations on the progress of genetics will not help.

If, in panic, our society should curtail fundamental genetic research, we would pay a huge price. We would slow our current progress in recognizing defective genes and preventing their spread; and we would block the possibility of learning to repair genetic defects. The sacrifice would be even greater in the field of cancer: for we are on the threshold of a revolutionary improvement in the control of these malignant hereditary changes in somatic cells, and this achievement will depend on the same fundamental research that also contributes toward the possibilities of cloning and of gene transfer in man. Finally, it is hardly necessary to note the long and continuing record of nonmedical benefits from genetics, including increased production and improved quality of livestock and crops, steadier production based on resistance to infections, vastly increased yields in antibiotic and other industrial fermentations, and, far from least, the pride that mankind can feel in one of its most imaginative and creative cultural achievements: understanding of some of the most fundamental aspects of the living world, including ourselves.

While specific curtailment of genetic research thus seeems impossible to justify, we should also consider briefly the broader proposal (see, for example, note 8) that we may have to limit the rate of progress of science in general, if we wish to prevent new powers from developing faster than we can provide an institutional framework to handle them. While one can hardly deny that this argument may be valid in the abstract, its application to our present situation seems to me dangerous. No basis is yet in sight for calculating an optimal rate of scientific advance. Moreover, only recently have we become generally aware of the need to assess and control the true social and environmental costs of various uses of technology. Recognition of a problem is the first step toward its solution, and now that we have taken this step it would seem reasonable to assume, until proved otherwise, that further scientific advance can contribute to the solutions faster than it will expand the problems.

Another consideration is that we cannot destroy the knowledge we already have, despite its potential for abuse. Nor can we unlearn the scientific method, which is available for all who wish to wrest secrets from nature. So if we should choose to curtail research in various fundamental areas, out of fear of possible long-range application, we must recognize that other societies may make a different choice. Knowledge is power, and power can be used for good or for evil; and, since the genie that brings new knowledge is already out of the bottle, we must learn to direct the use of the resulting power rather than curse the genie or try to confine him.

We cannot see how far the use of science as a scapegoat for many of our social problems will extend. But the gravity of the threat may be underscored by recalling that another politically based attack on science, Lysenkoism, utterly destroyed genetics in the Soviet Union and seriously crippled agriculture, from 1935 to 1965.[13] [This development illustrates ironically the unstable relation between political and scientific ideas: for Karl Marx had unsuccessfully requested permission to dedicate the second volume of *Das Kapital* to Charles Darwin![14]] Moreover, the current attacks on genetics from the New Left can build on, and have no doubt contributed to, widespread public anxiety concerning gene technology. Thus while a recent report prepared for the American Friends Service Committee[15] presents an open and thoughtful view on such questions as contraception, abortion, and prolongation of the period of dying, it is altogether opposed to any attempted genetic intervention, including the cure of hereditary disease.

Genetics will surely survive the current attacks, just as it survived attacks from the Communist Party in Moscow and from fundamentalists in Tennessee. But meanwhile if we wish to avert the danger of some degree of Lysenkoism in our country we may have to defend vigorously the value of objective and verifiable knowledge, especially when it comes into conflict with political, theological, or sociological dogmas.

Notes and References

1. P. B. Medawar, *The Future of Man* (Basic Books, New York, 1960); symposium on "Evolution and Man's Progress," *Daedalus* (Summer, 1961); G. Wolstenholme, (ed.), *Man and His Future* (Little, Brown, Boston, 1963); J. Lederberg, *Nature* 198 (1963):428; J. S. Huxley, *Essays of a Humanist* (Harper and Row, New York, 1964); T. M. Sonneborn, (ed.), *The Control of Human Heredity and Evolution* (Macmillan, New York, 1965); R. D. Hotchkiss, *Journal of Heredity* 56 (1965):197; J. D. Roslansky, (ed.), *Genetics and the Future of Man* (Appleton-Century-Crofts, New York, 1966); N. H. Horowitz, *Perspectives in Biology and Medicine* 9 (1966):349.
2. The term "genetic engineering" seemed at first to be a convenient designation for applied molecular and cellular genetics. However, I agree with J. Lederberg (*The New York Times*, Letters to the editor, 26 September [1970]) that the overtones of this phrase are undesirable.
3. Editorials, *Nature* 224 (1969):834, 1241; J. Shapiro, L. Eron, J. Beckwith, *Nature* 224 (1969):1337.
4. J. Beckwith, *Bacteriological Reviews* 34 (1970):222.
5. P. Handler, *Federal Proceedings* 29 (1970):1089.
6. J. Shapiro, L. MacHattie, L. Eron, G. Ihler, K. Ippen, J. Beckwith, *Nature* 224 (1969):768.
7. K. L. Agarwal, and 12 others, *Nature* 227 (1970):27.
8. S. E. Luria, in *The Control of Human Heredity and Evolution*, T. M. Sonneborn, ed. (Macmillan, New York, 1965), p. 1.
9. R. Briggs and T. J. King, in *The Cell*, J. Brachet and A. E. Mirsky, eds. (Academic Press, New York, 1959), vol. 1; J. B. Gurdon and H. R. Woodward, *Biological Reviews* 43 (1968):244.
10. B. Mintz, *Journal of Experimental Zoology* 157 (1964):85, 273.
11. A. Etzioni, *Science* 161 (1968):1107.
12. J. Lederberg, *American Naturalist* 100 (1966):519.
13. Z. A. Medvedev, *The Rise and Fall of T. D. Lysenko* (Columbia University Press, New York, 1969).
14. T. Dobzhansky, *Mankind Evolving* (Yale University Press, 1962), p. 132.
15. *Who Shall Live?* Report prepared for the American Friends Service Committee (Hill and Wang, New York, 1970).

33

Genetic Engineering: How Great Is the Danger?

This article appeared shortly before the controversy over recombinant DNA heated up. It focuses on a few key points that are expanded in later articles of this section.

Public concern over the potential dangers of genetic engineering in man now seems likely to be activated again, since a recent statement of a committee of the National Academy of Sciences[1] has brought to public attention the definite dangers of genetic engineering in bacteria.

Two major categories of genetic engineering in man may be envisaged. One, aimed at replacing defective genes, has given rise to fear that the technique would be used not only to cure disease but also to modify peoples' natures. Indeed, the prospect of parents shopping in a genetic supermarket, or of a tyrant specifying the genes in his subjects, would be harrowing. But for a realistic assessment of these dangers the distinction between single-gene traits and polygenic traits is crucial. The former depend on a single definable gene, with a recognizable qualitative effect (for example, the presence or absence of particular protein, such as sickle cell hemoglobin). In contrast, polygenic traits (for example, size and shape, strength and dexterity, intelligence and special talents, features of temperament), which are socially much more interesting, show a continuous range of variation, because they depend on the sum of the small contributions of many genes interacting with many environmental factors.

Editorial in *Science* 186 (1974):309.

The contrast in our knowledge of these two classes of traits is enormous. The success of molecular genetics has been confined to single-gene traits. For any behavioral trait we know only that many genes are involved: we have no idea how their products contribute to the circuitry of the 10 billion cells of the developing human brain. Moreover, we cannot identify *one* gene or protein whose variation contributes to the normal range of behavior, though we would need such information for many genes before we could try to modify behavior by manipulating DNA.

This vast ignorance about polygenic traits protects us against the main possibilities of harm from gene replacements. On the other hand, the possibilities for good are enormous, with increasing recognition of single genes that influence many aspects of man's health (such as specific immune responses). Hence it would be tragic to discourage efforts to overcome the technical obstacles—and these are still large.

The other major category of gene manipulation is the production of an exact gene copy of an individual. Such cloning, already accomplished with frogs, seems likely to become feasible in mammals fairly soon, and in a world facing severe food shortages the incentive to clone prize cattle will be strong. Extension to humans would indeed have grave and novel moral implications. But the dangers are hardly terrifying. If human cloning becomes feasible, and if it is then proscribed, an occasional violation would not shake the heavens. Moreover, if a tyrant wished to develop a particular kind of population he would not need cloning but could employ selective breeding, as used in animal husbandry since neolithic times.

Genetic engineering presents quite different problems in man and in bacteria. With bacteria the moral issues are simple, and they face us now. With man the moral issues are novel, and the problem is still in the future. But since we cannot predict when a particular kind of manipulation may become feasible, and since moral standards and social needs change with time, it would be presumptuous for us to try to guide future generations by our present wisdom.

It seems important for scientists to help the public to sort out these complex issues and avoid anxiety over improbable or distant developments. Such anxiety could lead to pruning of valuable major limbs on the tree of knowledge, rather than of branches with dangerous fruit.

REFERENCE

1. P. Berg, D. Baltimore, H. W. Boyer, S. N. Cohen, R. W. Davis, D. S. Hogness, D. Nathans, R. Roblin, J. D. Watson, S. Weissman, N. D. Zinder, *Science* 185 (1974):303.

34

Evolution, Epidemiology, and Recombinant DNA

By the time this letter was published the recombinant DNA debate had reached a high temperature. In an effort to defuse some of the wilder scenarios, I pointed out that even though the problem arose from discoveries in molecular biology, it was actually one in evolutionary biology, epidemiology, and bacterial physiology. It was therefore important to recognize that well-established principles in these fields provided solid grounds for believing that the recombinant bacteria under discussion were not as novel, and also not as likely to be dangerous, as the molecular biologists had at first assumed.

Of course, this statement did not stem the tide of public anxiety. Later subsidence, and the relaxation of the National Institutes of Health (NIH) Guidelines arose primarily from the fact that the expanding work with recombinant organisms failed to produce any illness. In addition, experiments showed that the bacterial strain used in all the work at that time was not well adapted to spread in nature, for after ingestion by volunteers in large amounts it disappeared rapidly from their intestines. However, neither of these empirical approaches can guarantee that the next novel recombinant might not survive and spread. The principles of evolutionary biology and epidemiology, emphasized in this letter, thus still seem to be our most valuable guide for estimating the dangers from future novel recombinants.

In attempting to assess the hazards of incorporating eukaryotic DNA into bacteria it is not enough simply to set up hypothetical scenarios: we

---

Science 193 (1976):442.

must also try to judge critically the underlying assumptions. The first assumption is that these experiments will breach an ancient barrier between eukaryotes [higher organisms] and prokaryotes [bacteria] and will thereby produce a radically novel class of organisms.

Principles from evolution and bacterial ecology offer our best guides for judgment. Bacteria in nature have long been exposed to DNA from lysed mammalian cells—for example, in the gut and in decomposing corpses. *Escherichia coli* can take up DNA after damage to the cell envelope, and one would expect random phenotypic variation to produce such damage occasionally (perhaps at frequencies of 10^{-5} to 10^{-10}). Homologous DNA is efficiently incorporated after entry, because its potential pairing with long regions of host cell DNA facilitates enzymatic crossover. Indeed, genetic recombination between bacteria (transformation) has even been observed in the human host. Incorporation of nonhomologous DNA is much less efficient but nevertheless can occur, presumably by transient pairing between adventitious short regions of complementarity. For example, deletions based on such "illegitimate recombination" occur at frequencies of about 10^{-9}.

With such low frequencies of both entry and incorporation, one could not expect to demonstrate natural hybridization between bacteria and man. Nevertheless, its scale almost certainly compensates for its inefficiency. Every person's gut is a huge chemostat, and the total population excretes about 10^{22} bacteria per day. Hence over the past 10^{6} years human-bacterial hybrids are exceedingly likely to have already appeared and been tested in the crucible of natural selection. If so, experimental DNA recombination will not be yielding a totally novel class of organisms.

A second assumption is that some of the recombinant strains are likely to spread and cause epidemics. Evolutionary principles are again pertinent. Nature selects for genetic balance: the contribution of a gene to Darwinian fitness depends on the rest of the genome. In bacteria, specifically, the introduction of a substantial block of foreign DNA would almost always lower the growth rate. With the short generation time of bacteria such a difference would lead to rapid outgrowth by competitors (unless the introduced genes promoted adaptation to alterations in the environment, such as the wide use of an antibiotic).

This argument is reinforced by a large body of epidemiological and experimental evidence. To cause communicable disease a potentially pathogenic organism must be able to survive in nature, in competition with other strains. It must also be able to be transmitted to a host, reach a susceptible tissue, and express its toxic potentialities there. Much current anxiety seems to be based on unawareness that microbial pathogenicity and communicability are complex and depend on a balanced genome. *Escherichia coli* carrying a gene for diphtheria toxin would be poorly suited to cause a diphtheria epidemic.

While bacteria carrying mammalian genes are thus unlikely to menace the public health, the risk of laboratory infection is much larger, since a heavy infecting dose of even a poorly communicable organism can cause disease in an individual. But this danger resembles that encountered with known pathogens, and it can be minimized by similar means. Perhaps the most valuable outcome of the current debate would be the requirement that those working on recombinant DNA be trained and supervised like medical bacteriologists.

I conclude that the risks in research on recombinant DNA require reasonable precautions but do not warrant public anxiety. A greater danger may be that the presumed analogy to nuclear weapons will lead to demands for virtually absolute freedom from risk. Yet the analogy to our mastery over infectious diseases is more apt. And if this field had faced similar demands, from its start, we might still be losing one-quarter of our children to communicable diseases. Is the balance of risk and benefit in research on recombinant DNA so much more unfavorable?

35

The Hazards of Recombinant DNA

This article expands the arguments offered in the preceding letter. It appeared in a debate with Robert Sinsheimer, who was the most responsible and thoughtful opponent of research on recombinant DNA. His views, originally presented on the same pages, were based primarily on the conviction that exchange of genes between prokaryotes (bacteria) and eukaryotes (higher organisms) was fraught with unforeseeable and dangerous long-term consequences for the biosphere. I could not see any reasonable grounds for that conviction, and it has since been strongly contradicted by the discovery that DNA is transferred between these kingdoms in nature.

Discussion of the potential risks of research on recombinant DNA has centered on the possibility of creating novel strains that might be a menace to public health: that is, they not only could cause disease in an individual but could spread in the human population. Since we lack the data required for reliably estimating that risk, we must depend on general principles. I would focus on two evolutionary considerations that have received relatively little attention: the degree of novelty of prokaryotes carrying eukaryotic DNA, and the requirements for effective spread of a pathogenic organism in nature.

With respect to the first, I find it presumptuous to think that man is creating an entirely new class of organisms by inserting eukaryotic DNA into prokaryotes. I also find no plausible scientific basis for

---

Trends in Biochemical Sciences, August 1976, pp. N178–180.

the suggestion that there is a natural barrier between the two groups which we breach at our peril. One can hardly doubt that human DNA (and DNA of human viruses) has always been getting incorporated into bacteria at a low rate. Bacteria are inevitably exposed to fragments of DNA from lysed mammalian cells—in the gut, on body surfaces, and massively in decomposing corpses. The entry of such fragments varies widely with circumstances and with the organism. In many gram-positive bacteria this entry is quite efficient, and genetic recombination by this mechanism (transformation) has been observed in the human host. With gram-negative bacteria (the predominant organisms of the gut) an extra layer in the cell envelope impedes DNA uptake, but if the barrier is damaged in appropriate ways transformation can be readily achieved (as in current recombinant DNA experiments). One would expect random spontaneous phenotypic variation also to produce the necessary damage occasionally in gram-negative cells (?10^{-5}, 10^{-10}).

After entering the bacterial cell, homologous DNA is efficiently incorporated, since the presence of long regions of complementary bases facilitates the pairing that precedes enzymatic crossover. The incorporation of nonhomologous DNA is much less efficient. It nevertheless does occur, presumably on the basis of transient pairing between short regions of complementarity. For example, this process (illegitimate recombination) is the basis for the incorporation of certain viruses into more or less random sites on bacterial chromosomes, and for deletion and insertion mutations, with frequencies of 10^{-6} to 10^{-9}.

The product of a low frequency of entry and a low frequency of incorporation would be too low to measure experimentally. The theoretical argument also does not lead to a reliable number. Nevertheless, the scale of the process in nature is almost certainly enough to compensate for its inefficiency. Every person's gut is a huge chemostat, excreting about 10^{13} bacteria per day, and so the output by the global population is about 10^{22} bacteria per day. It is therefore exceedingly likely that the kinds of recombinants we may produce in the laboratory are not altogether novel but have already been tested in the crucible of natural selection. (Indeed, since illegitimate crossing over is favored by short regions of homology, such as those provided by restriction sequences, past recombinations in nature may have occurred preferentially at the very same restriction sequences that are the sites of enzymatic recombination in the current manufacture of hybrid DNA in vitro.)

Of course, when investigators stitch together blocks of DNA from different sources they may produce combinations that could not have arisen in nature in one step, and so these products are less likely to have appeared in the past; but any estimate of increased hazard would have to depend on the details of the specific case.

Let us now look at the second evolutionary consideration: the Darwinian fitness, in nature, of a bacterial strain that has had a block of eukaryotic DNA inserted in its genome. What is its chance of competing successfully in the gut and being transmitted to other people? Here the underlying principle is clear. As Dobzhansky has emphasized, nature selects for balanced genomes: the evolutionary value of a gene cannot be weighed in isolation but depends on the genetic background in which it resides. Any particular small mutation, occurring in a genome that is already well adapted, has an exceedingly small probability of being advantageous. A large change, such as the introduction of a substantial block of foreign DNA, would almost certainly be disadvantageous. There is one important exception: a radical change in the environment (e.g., the widespread use of an antibiotic) will severely impair the fitness of previously adapted organisms and will favor progeny that have picked up an appropriate (drug-resistant) gene. But even here the probability of our producing truly novel combinations in the enteric group is exceedingly low. The enteric bacteria have been exchanging resistance genes on plasmids for a long time, even before the advent of antibiotic therapy, and the combinations that have thus far been identified can be only a minuscule fraction of the combinations that exist at present.

This evolutionary argument for the complexity and subtlety of bacterial virulence is reinforced by a large body of experimental evidence. To cause communicable disease an organism must be able not only to have a toxic effect but to survive somewhere in nature, to be transmitted to a host, to reach a susceptible tissue, and to express its toxic potentialities there. Much current anxiety seems to be based on a failure to appreciate the complexity of bacterial virulence. For example, the production of a particular toxin may be essential for the virulence of a particular organism, but adding the gene for that toxin to quite a different organism is not sufficient to confer effective virulence. Thus, *Escherichia coli* has no doubt frequently exchanged genes with the tetanus bacillus, which is also a common inhabitant of the mammalian gut, yet *E. coli* strains producing tetanus toxin have not been detected. Indeed, the insertion of extra genes will ordinarily give an organism a "genetic load" that decreases its ability to multiply in nature in competition with other organisms. And since bacteria have a short generation time even a slight differential in growth rate will cause the less effective competitor in a mixed population to be rapidly outgrown.

While the risk to the public health from the incorporation of mammalian DNA into bacteria thus does not seem substantial, the risk of laboratory infection may be much larger. For even if an organism is poorly communicable, a heavy infecting dose in an individual can cause disease. But this danger is no different from that encountered in laboratory work with known pathogens. It can be avoided if molecular geneticists working with recombinants learn, and take seriously, the precautions that have become standard in medical bacteriology laboratories.

These evolutionary and epidemiological considerations suggest that the risks in responsible research on recombinant DNA are not likely to be even as great as those that investigators have created in the past in the course of discovering and learning to control the agents of infectious disease. The present risks are real, and they require reasonable precautions, but I do not believe they are so threatening as to justify restrictions that would seriously hamper valuable research, or to warrant public anxiety.

The main danger, as I see it, is neither the production of novel, dangerous pathogens nor the imposition of regulations that would increase the cost of research in this field. It is the encouragement of demands for virtually absolute freedom from risk, from a public that fears a biological analogy to the development of nuclear weapons. This analogy is weak. Our mastery over most infectious diseases is a much more apt precedent, and it could not have been achieved if earlier microbiologists had been faced with the present demands for freedom from risk. Research on pathogenic microbes has cost the lives of several hundred investigators and has saved millions of other lives. Is the balance of risk and benefit in research on recombinant DNA so much more unfavorable?

36

Debate on Recombinant DNA Research

At the height of the public debate over the dangers of research on recombinant DNA the news magazine of the American Chemical Society arranged for an exchange among three people with different points of view: Erwin Chargaff, a brilliant biochemist and a writer with a sharp pen, who had long questioned the value and excoriated the style of molecular biology; Sheldon Krimsky, a social scientist who was interested in promoting direct public involvement in all decisions that affect their interests; and me. Each of us submitted a paper and then commented on the other two. My paper is not reprinted here, since it covers ground that has already appeared in this volume. However, my comments on the other papers, presented below, contain additional responses and may give some of the flavor of the controversy at that time. A review of Krimsky's recent work on genetics appears as number 38.

Professor Chargaff is an old friend and colleague, and I have long admired his brilliant literary style and his fundamental contributions to science. Moreover, I believe the original Watson-Crick paper did not adequately acknowledge the role that his base ratios played in the elucidation of the structure of DNA. It is therefore painful to have to rebut Chargaff's article. It is all the more painful since his philosophical remarks do not lend themselves to point-by-point analysis, and so I will have to comment on his judgment. But the issue is too serious to permit a courteous evasion.

Chemical and Engineering News (May 30, 1977):42.

In the first paragraph, Chargaff makes what seems to me to be his most significant point: "we have always underestimated the specificity, the exquisite fit, of all life processes." This statement appears to recognize the importance of what I have called, in more prosaic and technical terms, a balanced genome. I would therefore have expected him to come to the same conclusion that I have emphasized: that the jerry-built hybrids resulting from molecular recombinations have very little chance of yielding dangerous new organisms endowed with evolutionary fitness.

Unfortunately, he does not draw this conclusion, nor does he try to apply relevant scientific principles to the problem at hand. What he presents seems instead to be one more item in the long record of his feud with molecular biologists—a group that he characterized many years ago as biochemists practicing without a license.

I would not take issue with Chargaff's condemnation of the brash style of many of the practitioners of molecular genetics. But his extraordinarily negative views on the validity and the significance of the content of this field are another matter, and the relevance of these views to his assessment of recombinant DNA research cannot be ignored.

It is sad that a man with such a strong sense of history would approach molecular biology with so much emphasis on its methodological imperfections and its bold inferences. How would he have reacted a century ago, when a respected chemist named Louis Pasteur began speculating loosely about the possibility that invisible microbes might be causing diseases in man as well as in wine?

And a sense of history also seems to be suppressed when Chargaff joins the handful of scientists who are opposing research on recombinant DNA. Being so fearful of a purely conjectural risk, would they have approved when Robert Koch in Berlin, in 1882, identified the agent of tuberculosis—and then began to grow this known dangerous organism in quantities large enough to infect all the inhabitants of the city?

Chargaff's hyperboles and shafts of wit have long entertained and stimulated the scientific community. But they take on a new significance when they address an issue that frightens the general public.

At the recent annual meeting of the National Academy of Sciences, President Philip Handler's report, and a resolution passed by the membership, emphasized the danger that pending legislation could seriously impede the valuable explorations made possible by the new recombinant DNA technology. Moreover, if this precedent succeeds and persists, it will surely extend to other areas of scientific research. Would Chargaff like to be a sorcerer's apprentice, starting a flood of vulgarization and bureaucratization of science and then wringing his hands?

With respect to Professor Krimsky's article, the citizens' committee of which he was a member started with suspicion of the ability of scientists to regulate themselves responsibly, and ended up with a unanimous vote of confidence in the main features of the National Institutes of Health guidelines. For their patience, restraint, and objectivity in listening to many dozens of hours of testimony in a highly charged atmosphere, they deserve strong commendation.

They also added a provision for local supervision of adherence to the guidelines, which seems reasonable. However, I would question the wisdom of their further initiative in shifting one set of experiments from P2 to P3. This decision sets a precedent for a group of laymen to go beyond formulating general policy on a highly technical issue.

We are fortunate that the Cambridge review board did so well, but we certainly cannot expect the same of all local groups. If the pending federal legislation should include local option, the restraints are likely to escalate. Such excessive and variable restrictions would make the research more expensive or even impossible, and they would result in disruptive migration of biologists.

Since I believe the underlying anxieties are not warranted, I would predict that some years later we would ask—as in the quarantining of lunar astronauts—why so many tens of millions of dollars had been wasted. But in this case the losses would go far beyond waste of money, and could threaten the spirit of the scientific enterprise.

Let me close by correcting the impression, widespread among concerned laymen, that the experts are in sharp disagreement over the hazards of research in this field. The problem is to recognize who are the experts. They are the specialists in epidemiology and infectious disease. In fact, these experts do not perceive the hazard as a serious one, as was clear in a recent "Nova" program on public television (but unfortunately was not emphasized or documented). I can report the same response from my own contacts.

Substantial input from this field will clarify much that has been overlooked in the debate. Meanwhile, the record of laboratory work on dangerous organisms and our limited capacity to influence evolution by genetic manipulation—both summarized in my paper—do not justify the alarm exhibited by Chargaff and Krimsky.

37

Epidemiological and Evolutionary Aspects of Research on Recombinant DNA

This piece was presented at a forum, organized by the National Academy of Sciences, that was intended to clarify the hazards of this new technology and to raise the quality of the ongoing debate. Much the same material was covered in a paper published in American Scientist *65 (1977): 547–555, entitled "The Recombinant DNA Scenarios: Andromeda Strain, Chimera, and Golem."*

In earlier times the academy would no doubt have responded to such public concern over a scientific issue by setting up a blue-ribbon commission in which respected senior statesmen would hear testimony on all sides and then submit a judicious report. It is an index of the times that the academy adopted instead quite a different procedure: a forum open to public participation, with a proponent and an opponent on each topic. This approach created a tense, politically charged atmosphere. Moreover, in a scene never expected in the halls of the academy, a group led by Rifkin and Howard (whose book Who Should Play God? *I have discussed in the first comment in this section) unobtrusively occupied the perimeter of the gathering audience, and the moment the meeting was opened they unfurled banners and chanted until allowed to make a half-hour speech—one that, as the chairman aptly said, "pushed at an open door."*

My reaction at the time was that the format of this meeting gave too much exposure to the handful of scientists with demagogic or idiosyncratic positions on this issue, instead of restoring the kind of atmosphere that could produce a responsible assessment. On the other hand, the members of the press were able to compare the quality of the arguments on both sides and to witness the chilling take-over of the chamber; the experience may have helped them to better decide which appraisals were closest to reality.

Forum on Recombinant DNA (Washington, D.C.: National Academy of Sciences Press, 1977).

Several charges have been leveled against proponents of research on recombinant DNA: selfishness, in risking the production of an Andromeda strain in order to satisfy their curiosity; blasphemy, in meddling with evolution; and irresponsibility, in bringing us closer to genetic engineering in man. These charges have been based on the assumption that we are entirely in the dark in trying to assess these dangers. But this is not so. On the question of the hazard of an epidemic a good deal of pertinent theoretical and factual information is available from the science of epidemiology (concerned with the genetic and the ecological factors that influence the spread of disease), and from evolutionary theory (of which epidemiology may be viewed as an applied branch). Evolutionary theory also has serious implications for the more long-range danger of possibly fouling up evolution. This paper will review some of the relevant information, concentrating on the risk of producing an epidemic, and considering this problem in terms of three component risks: that a harmful organism may inadvertently be produced; that it may cause a laboratory infection; and that it may spread into the community.

In approaching the subject from this perspective I would like to express my agreement with Dr. Jonathan King on one point: that the Asilomar conference did not have sufficient input from experts in infectious disease. I further regret that this field continues to be relatively neglected in the current discussion. For since we are dealing more with a problem in epidemiology than with one in molecular biology, epidemiological principles provide the most reasonable basis for present estimates of risk. Moreover, though the risk of an epidemic will ultimately have to be assessed in terms of future experience with various recombinants, even the most favorable experience will not eliminate the specter of a future Andromeda strain unless we interpret it in terms of epidemiological principles.

Underlying Principles

Natural Selection

Evolutionary change arises ultimately from hereditary variation, but its direction is dominated by natural selection. It is dramatic for George Wald to state that research with recombinants is dangerous because "a living organism is forever"—but a more balanced statement would also note that only an infinitesimal fraction of the products of evolutionary experimentation survive, the rest being ruthlessly culled out by natural selection. In particular, within a species the process of sexual reproduction produces a virtually infinite variety of recombinants, among which the standard pattern of selection is a stabilizing (normalizing) one: excessive deviations from the norm make an organism less effective in the Darwinian competition. It is only when the environment is altered that certain deviants from the norm turn out to be better adapted to the new environment, and selection then becomes directional.

It should also be emphasized that *all natural selection is for a balanced genome.* A gene that increases or decreases a trait is selected for not in a vacuum, but only if it is coadapted to the rest of the organism's total set of genes.

The Meaning of Species

As evolution proceeded from prokaryotes to eukaryotes it created the mechanism of sexual reproduction. By reassorting the genes of paired parents this process provides vastly increased genetic diversity for natural selection to act on. But since a successful organism must have a reasonably balanced set of genes the production of unlimited recombinations from the total pool of genetic material in the living world would not be useful. Hence the development of sexual reproduction was accompanied by the development of species: groups of organisms that reproduce in nature only by mating with other members of the same group, and not with members of other species. The evolutionary value of such fertility barriers between species is clear: to avoid useless production of grossly unfit, nonviable progeny.

Bacterial Genetics

Though Darwin was unaware of the existence of the invisible world of microbes, their slow absorption into the Darwinian framework began, unwittingly, with Pasteur's demonstration that different media, such as milk or grape juice, select for different organisms from the same mixture of contaminants that can reach them from the air. But it was not until the 1940s that heredity in bacteria was shown to depend, as in higher organisms, on a set of genes, linked on a chromosome and capable of mutation, transfer, and recombination. Indeed, with this development it became possible to use microbes to demonstrate the force of natural selection in an overnight experiment. In addition, with the emergence of molecular genetics from microbial genetics it became possible to provide the ultimate proof, from DNA sequences, for a crucial prediction of modern evolutionary theory: that the accumulation of changes in genes is the basis for the divergence of organisms in evolution.

Unlike eukaryotes, prokaryotes ordinarily reproduce by asexual cell division, which means that the genetic properties of a strain remain constant for generation after generation, except for rare mutations or for rare transfers of a block of genes from one cell to another. These gene transfers, which are usually mediated by plasmids or viruses, do not show a sharp species boundary: they simply become less efficient the greater the evolutionary separation between the donor and the recipient. Prokaryotes therefore have no true species. *E. coli*, for example, is the name given to a range of strains with certain common features and also with a variety of differences—in surface molecules, nutrition, growth rate, sensitivity to inhibitors, etc. These differences determine the relative Darwinian fitness of various strains for various environments.

Bacterial Ecology

Every living species is adapted to a given range of habitats. The set of bacterial strains called *E. coli*, and such closely related pathogens as the typhoid and the dysentery bacillus, thrive only in the vertebrate gut. In water they survive temporarily but quickly die out. (Indeed, for that reason the *E. coli* count of a pond is a reliable index of its continuing fecal contamination.) In the gut there is intense Darwinian competition between strains, depending on such variables as growth rate, nutritional requirements, ability to scavenge limited food supplies, adherence to the gut lining, and resistance to antimicrobial factors in the host. Hence most novel strains are quickly extinguished, in the kind of competition envisaged by Darwin for higher organisms. With bacteria the process is very rapid, because the generation time is as short as twenty minutes and the selection pressures are often intense.

It is easy to demonstrate that the environment in the gut (i.e., type of food and physiological state) plays a decisive role in determining the distribution of organisms in its normal flora. For example, when a baby shifts from breast feeding to solid food the character of the stool changes dramatically, as lactic acid bacteria, which produce sweet-smelling products, are replaced by *E. coli* and other foul organisms. Moreover, efforts to reverse the process in adults, by administering large numbers of lactic acid bacteria in the form of yogurt, have not been successful.

Pathogenicity

Various kinds of infectious bacteria differ from each other in several distinct respects: *infectivity* (i.e., the infectious dose, ranging from a few cells of the tularemia bacillus to around 10^6 cells of the cholera vibrio); specific *distribution* of the organisms in the body; *virulence* (i.e., the severity of the disease once the infection has overcome natural resistance); and *communicability* from one individual to another (including length of survival in nature). Each of these attributes, like any complex property, depends on the coordinate, balanced activity of many genes, capable of independent variation.

It is especially important to distinguish the ability to *produce* a serious disease from the ability to *spread*. For example, the tetanus bacillus produces a powerful toxin, but it is a normal, noninvasive inhabitant of the gut: it can cause fatal illness only when it gains access (usually by trauma) to a susceptible tissue, and so a patient with tetanus is not a menace to his contacts.

Estimation of the Hazards

In turning now to the risks, I would note that they are often not as directly commensurable with benefits (i.e., expressible in similar units) as are costs compared with benefits. For this reason a particular risk must be judged for acceptability not only in terms of a comparison with benefits but also in terms of its probable increment to the related risks that we already live with. I would further emphasize that it is easy to draw up scary hypothetical scenarios if one's imagination need not be limited by considerations of probability. But any realistic discussion must consider probabilities. And as I mentioned earlier, we must consider three probabilities: that experiments with a given kind of DNA will produce a dangerous organism, that that organism will infect a laboratory worker, and that the organism will escape and spread in the community or the environment.

Risk of Producing a Harmful Organism

There is no doubt that molecular recombination in vitro could produce pathogenic derivatives of *E. coli*. For example, if a strain carrying the gene for a potent bacterial toxin multiplied enough in the host, or even if it could not multiply but were taken up in a large enough dose, it could cause disease. A strain carrying a tumor virus might also be hazardous. However, its production of a pathogenic effect is less certain. For unlike a toxin producer, such strains would require for pathogenicity more than the normal function of the foreign DNA within the bacterial carrier: it would require release of that DNA from the bacterial cell and its infection of animal host cells. While that probability may be very low, we cannot assume that it is negligible. Both these kinds of strains are appropriately prohibited in the National Institutes of Health guidelines today.

I would like to concentrate on a kind of experiment that is allowed, but that is causing great concern and is restricted to P3 facilities: the so-called "shotgun" experiment, in which one transfers random fragments of DNA from mammalian cells. Two considerations convince me that the danger in such experiments has been enormously exaggerated.

First, such cells have a million gene equivalents, and since each recombinant strain would contain only a few genes, the probability of isolating a strain with genes for a toxic product or for a tumor virus is exceedingly low. Second, I would seriously question whether the novelty that we fear in the products of such experiments is real.

The reasons for this doubt are the following. it is known that bacteria can take up naked DNA from solution. In fact, two different strains of pneumococcus have been shown to be able to produce a third, recombinant strain in an animal body, by release of DNA from a lysed cell of one strain and its uptake by an intact cell of the other. Moreover, in the gut bacteria are constantly exposed to fragments of host DNA, released by death of the cells lining the gut; while bacteria growing in carcasses have a veritable feast. To be sure, the efficiency of uptake of DNA by bacteria (especially the kinds found in the gut) is very low; but on the other hand, the scale of the exposure in nature is extraordinarily large—around 10^{20} bacteria are excreted collectively by the human species per day. Hence it seems virtually certain that recombinants of this general class have been formed innumerable times over millions of years and are being formed in nature today. If they had high survival value we would be recognizing short stretches of mammalian DNA in *E. coli*. We do not. On the other hand, naturally occurring recombinants might be appearing and even causing transient epidemics, which are escaping our attention. But then we would have to ask how much our laboratories could add, performing experiments on a scale of a billion times smaller.

Risk of Laboratory Infection

Having considered the probability of inadvertently producing a harmful organism, we must now consider the probability that such an organism would cause a laboratory infection. Let us assume the worst case, at present prohibited: an *E. coli* strain producing a potent toxin absorbable from the gut, such as botulinus toxin. The danger of harm from a laboratory infection with such a strain would be real. However, there are a number of reasons to expect it to be less than the danger encountered with the pathogens that are handled every day in medical laboratories.

(*a*) In the history of microbiology about 6,000 instances of laboratory infection have been recorded. Moreover, these cases were largely due to various agents of respiratory infection, spread by droplets; and the rate has dropped markedly since safety cabinets were introduced in the 1940s. In contrast to such respiratory infections, enteric infections arise through the swallowing of contaminated food or other material. Hence even the most virulent enteric pathogens are relatively safe to handle with simple precautions, such as not eating or smoking in the laboratory.

(*b*) Strain K12, used in almost all genetic work with *E. coli* (including current work with recombinant DNA), has been transferred in the laboratory for over fifty years, and during this time it has become well adapted to artificial media, at the cost of becoming deadapted to the human gut. In fact, in recent tests in man this strain disappeared from the stools within a few days after a large dose (much larger than what one would expect from a laboratory accident). Its problems of survival outside the laboratory are analogous to those of a delicate hothouse plant thrown out to compete with the weeds in a field.

(*c*) The addition of a block of foreign DNA to an organism will ordinarily decrease its adaptation to survival in nature. The contrary likelihood, of improving adaptation by such an insertion, is obviously all the smaller if the source of the DNA is distant in evolution from the recipient. A pertinent analogy here would be that of taking a specialized part from one kind of machine (e.g., an automobile) and expecting it to work well in a very different machine (e.g., a watch).

(*d*) A very large safety factor is added by the provision in the present guidelines for biological containment. All work with mammalian DNA must be carried out only in EK2 strains, which have a drastically impaired ability to multiply, or to transfer their plasmid, except under very special conditions provided in laboratory. The presently certified EK2 strain has several stable mutational defects (i.e., deletions) that prevent it from multiplying under the nutritional conditions of the gut. But the protection goes much farther, and reaches a degree that is unprecedented in the annals of man's exploration of potentially hazardous new materials: this material has been coded for self-destruction. For example, these mutant cells require diaminopimelate, a constituent of cell wall; and without it they can continue to grow and expand but cannot form more wall, and so they quickly burst. Accordingly, under conditions similar to those in the gut such an EK2 strain not only fails to multiply, but less than 1 in 10^8 cells survives after twenty-four hours—and it would be an extraordinarily sloppy laboratory accident that would result in ingestion of as many as 10^8 cells. In addition, while the cells are dying off in the absence of diaminopimelate they are severely impaired in their ability to transfer plasmids to other, well adapted cells—and this is the important point for the danger of spreading harmful genes. Finally, not only the cells but also the plasmids being used to carry recombinant genes are also weakened mutant derivatives, selected for severe impairment of their ability to be transmitted from the host cell to another cell.

We thus see that even with a strain known to carry the gene for a potent toxin the production of disease in a laboratory worker would require the compounding of two low probabilities: that the strain will initiate an infection; and that it will survive long enough to cause harm despite its disadvantages of being a laboratory-adapted strain, carrying the burden of foreign DNA, and carrying the very large burden of being a suicidal EK2 strain. With shotgun experiments we have a third, very low probability, already mentioned: that of having picked up a dangerous gene from normal mammalian tissue.

I conclude that with the kinds of recombinants now permitted the danger of a significant laboratory infection is vanishingly small compared with the dangers encountered every day by medical microbiologists working with virulent pathogens. And such dangers must ultimately be balanced against the potential benefits. In the United States, up to 1961, of the 2400 recorded cases of laboratory infections 107 were fatal—over half of these from diagnostic laboratories. Balancing this cost, millions of lives have undoubtedly been saved by bacteriological research and diagnosis.

On the other hand, even if the risks in recombinant DNA research are really small, it is important to keep all the probabilities low. Hence it is important for molecular biologists working in this area to learn, and to use, the standard techniques of medical microbiology. Indeed, the main benefit from the current discussion might well be the enforcement of such practices.

Risk of Spread

I now come to the most important point of all from the point of view of the public: the enormous difference between the danger of causing a laboratory infection and the further danger of unleashing an epidemic. Let us look at a few facts. In our government's bacteriological warfare laboratories at Fork Detrick, working for twenty-five years on the most communicable and virulent pathogens known, 423 laboratory infections were seen. Moreover, most of these infections occurred via respiratory transmission, over which control is very imperfect. Nevertheless, *only a single probable case of secondary spread* to a member of the family or to any person outside the laboratory was seen. Similarly, in the Center for Disease Control of the U.S. Public Health Service 150 laboratory infections were recorded, with only one case of transmission to a family member. Elsewhere in the world about two dozen laboratory-based microepidemics have been recorded—and each involved at most a few outsiders.

With enteric pathogens the danger of secondary cases is minimal, for with this class of agents modern sanitation provides infinitely better control than we can provide for respiratory infection: the appearance of a case of typhoid, in contrast to that of influenza, does not lead to an epidemic. Enteric epidemics appear only when sanitation is poor or has broken down, or when a symptom-free carrier with filthy personal habits serves as a food handler; and such epidemics are always small (except when sewage freely enters the water supply). Moreover, the focus of some critics on the debilitated or the young, as exceptionally susceptible victims, is not realistic: we are dealing with interruption of the chain of transmission, and not with wide spread of the organisms at a low density.

This information is clearly pertinent to recombinants in *E. coli*. For while widespread apprehension has arisen from the presumption that this procedure will produce biparental chimeras, with totally unknown properties, the fact is that the recombinants envisaged are all genetically 99.9 percent *E. coli*, with about 0.1 percent foreign DNA added. It is not conceivable that such an organism could have a radically expanded habitat, no longer confined to the gut. It is even harder to see that the organism would be more communicable, or more virulent, than our worst enteric pathogens, which cause typhoid or dysentery. The Andromeda strain remains entertaining science fiction.

I conclude that if by remote chance a recombinant strain should be pathogenic, and if it (or a recipient of its plasmid) should cause a laboratory infection, that infection would give an early warning. Moreover, if a case should appear outside the laboratory the enteric habitat of *E. coli*, combined with modern sanitation, provides powerful protection against the chain of transmission required for an epidemic.

Tumor viruses present a special problem. Unlike other viruses, they do not cause disease regularly after infection but require special circumstances. Indeed, it is their occasional presence in apparently normal animal tissues that has given rise to fear of "shotgun" experiments.

On the other hand, any conceivable infection by a bacterium containing a tumor virus genome would have a long latent period before disease could appear, and so we would lack the early warning that would be seen with a bacterium producing a potent toxin. However, this loss of one protective feature is balanced by the fact that viruses, by definition, have their own means of spread. Indeed, in general the natural spread of viruses is even more effective than that of bacteria, for each infected animal cell produces thousands of infectious virus particles, while each bacterium produces two daughter cells. Moreover, since viral DNA in a bacterium would have to get out of its host cell and get into human cells, through an extremely inefficient process, it is hard to imagine that that DNA in a bacterium would be more hazardous than that same DNA in its own infectious, viral coat, adapted by evolution for entering animal cells. Indeed, if we fear the danger of such indirect uptake of unrecognized tumor virus DNA from normal mammalian tissue, via a bacterial vector, we must ask whether the direct ingestion of such mammalian tissue, as in a "rare" steak, may not present at least as great a danger. Finally, if we fear that tumor viruses are sufficiently widespread to create a significant danger of being included in DNA fragments from normal tissue, we must ask how much that wide distribution could be increased by the remote chance of inadvertent further spread by the bacterial hybrids created by shotgun experiments.

I am not suggesting that we should be concerned about the danger of acquiring a cancer by eating rare meat (or by receiving a transfusion, which inevitably has a fair chance of coming from a person with an undetected early cancer). I am suggesting only that the danger of using recombinant DNA to study tumor viruses must be judged against that background, as well as against the background of the virus's own distribution and inherent ability to spread.

In the light of all these considerations, we must ask whether the danger of an epidemic really merits deep concern by the general public. To be sure, the problem of minimizing the risk of laboratory infections should concern those involved with such laboratories, just as with laboratories dealing with known pathogens. And I believe investigators have the right to take such risks for themselves, as they do daily in working with pathogens (including such unknowns as the agent of "Legionnaires' Disease"). But we have seen that by any reasonable analysis the risk of producing a serious epidemic with *E. coli* containing random fragments of mammalian DNA seems very much less than the risk from pathogens that are being cultivated in laboratories all the time. *I therefore see no realistic basis for public anxiety over this issue,* any more than over the way laboratory work on known pathogens is conducted.

The National Institutes of Health Guidelines

In the face of the alleged dangers that have been so vividly portrayed, I cannot blame the public for having a high level of anxiety. I also would regard the present guidelines as a reasonable response to that anxiety. On the other hand, in the light of the technical realities that I have discussed above I would regard these guidelines as excessively conservative. This is especially true of the experiments with mammalian DNA, which offer enormous promise in the analysis of the structure and the regulation of mammalian genes and in the manufacture of valuable human gene products.

The guidelines contain a provision for periodic revision; and since these revisions (or the nature of any future legislation) will depend on public attitudes as well as on the results of actual experience with the organisms, there is need for a great deal of public education, based on the relevant scientific facts and principles.

In this connection I would criticize the *New York Times* for the article by L. Cavalieri on recombinant DNA in its *Sunday Magazine* (August, 1976). Though the writer is a molecular biologist whose official credentials would lead the reader to expect a reasonable degree of objectivity, the article was inflammatory and it exhibited extraordinarily little understanding of either microbiology or evolution. In discussing *E. coli* as though it were a standard, uniformly distributed organism, which would carry with it through the world any additional genes that one might insert, the writer ignored the most important factor of all: natural selection among the innumerable strains of *E. coli.* He also made the remarkable statement that the insertion of tumor viruses into bacteria may make them infectious—as though viruses are not infectious. And he suggested that scientists working in this field may produce yet another Andromeda strain—as though the first strain existed in fact rather than in fancy.

Given the present level of public anxiety, scientists in this field seem quite willing to accept the guidelines. But I hope it will not be too long before these rules are modified in the light of further experience. For since the technique is potentially useful for a wide variety of problems, a requirement for excessively elaborate facilities will add up to a very large expense and will inevitably inhibit desirable experiments. The principle of erring on the side of caution is laudable up to a point—but if it is pushed too far it can end up being paralytic.

Intervention in Evolution

The Prokaryote-Eukaryote Barrier

The hazard that we have been discussing—that of creating novel, dangerous organisms—is a legitimate cause for public concern: there is no question about society's right to limit activities that may harm others. However, when we ask with Dr. Sinsheimer whether our increasing power to manipulate genetic material creates long-term evolutionary dangers we are in quite a different area, involving the concept of dangerous knowledge rather than dangerous actions. Perhaps we can clarify the issue by trying to translate into more specific terms some of the general sources of apprehension that Dr. Sinsheimer has expressed in various publications.

(*a*) He questions our moral right to breach the barrier between prokaryotes and eukaryotes, since we simply cannot foresee the consequences. This argument seems to turn evolutionary principles through 180 degrees. Evolution is concerned with selection for fitness, in the Darwinian sense. The barriers that it has established between species are designed to avoid wasteful matings, i.e., matings whose products would be monstrosities, unable to survive, rather than monsters, able to take over. Since survival of an organism depends upon a balanced genome it is not surprising that evolution proceeds in small steps, which will not excessively unbalance the genome in one respect while improving its adaptation in another. And since for this reason even closely related species cannot form hybrids in nature, it is exceedingly unlikely that artificial transfers of genes between the most distant organisms—man and prokaryotes—would pass the test of Darwinian fitness.

(*b*) "This is the beginning of synthetic biology." I wonder whether this statement can really be defended. Man has been meddling with evolution since neolithic times, domesticating animals and plants by selective breeding and also cloning and grafting plants.

(*c*) "We no longer have the absolute right of free inquiry." But we never had: visibly dangerous procedures have always been subject to social limitations. But to invoke dimly foreseen, undefined dangers as a basis for limitation seems to be starting on the slippery slope of excluding dangerous ideas rather than dangerous actions.

(*d*) A further push in this direction may be seen in the statement that power over nucleic acids, as over the atomic nucleus, "might drive us too swiftly toward some unseen chasm. . . . We should not thrust inquiry too far beyond our perception of its consequences." I would paraphrase this statement and suggest that we should not thrust our limitations on research too far beyond our perception of its hazards. Otherwise we will find ourselves reenacting the drama of Galileo and Urban VIII, and we will be trying to play the role of God (or of his representative). The analogy is uncomfortably close: for the mystical quality of the current argument suggests that at its core the issue is whether man's possible interference with evolution is not blasphemous.

Genetic Engineering in Man

Perhaps the most significant of Sinsheimer's statements is his suggestion that the study of recombinant DNA in bacteria is the beginning of a genetic engineering that will ultimately extend to man. Here, in contrast to the vagueness of the preceding propositions, we finally come to something concrete that one can wrestle with.

I would suggest that concern over genetic engineering in man is utterly irrelevant to the question of the danger of creating an epidemic; hence it is irrelevant to Sinsheimer's recommendation that all research on recombinant DNA be presently restricted to a few maximum security federal facilities. This concern also seems irrelevant to the question of breaching the prokaryote-eukaryote barrier; for while gene transfers across this border at the cellular level, in either direction, are of great scientific interest, it is hard to envisage any reason to try to introduce into man genetic material from the opposite end of the evolutionary spectrum. Yet vague concern over possible extensions of gene manipulation to man, even more than concern over epidemics or over meddling with evolution in general, may lie at the heart of much of the uneasiness over recombinant DNA research. And because of the enormous publicity given to our new power to splice blocks of DNA into plasmids, we have perhaps lost sight of the fact that this development is no more radical a step toward genetic engineering in man than are many other steps, which have aroused no such public terror. These include the isolation of a gene, its chemical synthesis, the cultivation of human cells, the use of viruses to incorporate genes into those cells, and the achievement of genetic recombination in vitro between human cells and other animal cells.

The prospects of genetic engineering in man received extensive discussion in 1970, which then subsided; and I see no reason to modify today the analysis that I published then (*Science* 170:1279), except to agree with Dr. Baltimore that replacement of bone marrow cells may no longer be very distant. However, since the question has been reactivated by the very different question of genetic engineering in bacteria, I would like to make a few brief points.

First, as far ahead as it is profitable to look, the medical aim of genetic engineering in man is simply gene therapy, for diseases due to defects in single genes with a well defined chemistry. (Cloning is another matter: its specific aim is to avoid genetic recombination, and its social purpose would not be medical.) For gene therapy of most hereditary diseases we would have to be able to introduce DNA in a reliable, controlled way, in the right cells: and I believe we are still a long way from that goal. But even if this guess is wrong, and if we succeed in genetically curing such diseases as phenylketonuria and cystic fibrosis, it is clear that we would still be very far from being able to manipulate in any useful way the large number of genes, all still undefined, that specifically direct the development and the function of the brain. Moreover, in a developed organism, with an already formed brain, no conceivable manipulation of DNA could reorganize the wiring diagram of that brain—which is surely the main basis for the genetic component of human behavioral diversity. Hence the possibility that a tyrant could use genetic engineering to manipulate personalities seems still too remote to justify present concern. Finally, even if we could use genetic technology in this way I would question whether the technological imperative would necessarily (or even likely) lead us to do so. For the simple but effective techniques of selective breeding and artifical insemination are already available, and yet they are not being used to influence the human gene pool.

Philosophical questions about the effects of science and technology on man's fate go back to Galileo—and the history of Italy's fate, in losing that early head start, should give us pause. For better or worse, we cannot unlearn the scientific method; and if we restrict it in one country it will turn up in another. To be sure, our world has only recently come to realize how large (and often unexpected) is the price for various aspects of technology, how finite our terrestrial resources, and how clumsy our responses to the need to limit the size of our population and its demands on those resources. Faced with these crushing problems, it is only too easy to take the benefits of science and technology for granted and to object to the new problems that they are raising. But in the long run it is difficult to see how we can plot a more prudent course than to continue to advance knowledge, while increasing our efforts to recognize (and to minimize) the hazards and the costs of its specific possible applications *as soon as they become visible.*

I share Sinsheimer's concern for the future, and his passionate advocacy of vigilance. But the vigilance must be directed at specific, definable applications of knowledge. Vigilance concerning new knowledge that *might* someday be misused is a threat to freedom of inquiry, and I believe a threat to human welfare. We may conceivably be entering dangerous territory in exploring recombinant DNA—but we are surely entering dangerous territory if we start to limit this exploration on the basis of our incapacity to foresee its consequences.

38

Along the Road to Asilomar

By 1982 the DNA controversy had died down, the research had continued to expand without causing any illness, and the NIH guidelines had been relaxed. It was therefore a good time for someone to write a history of this remarkable episode in the relations between science and society. Sheldon Krimsky was an appropriate candidate. He had been a member of the Citizens' Review Board that dealt with the issue in Cambridge, Massachusetts, and this experience led him, as a social scientist, to develop a professional interest in the problem.

Unfortunately, Krimsky's distrust of elites (including those defined by their special knowledge of a technical subject), and his confidence in participatory democracy, continued to guide his interpretation of this history. Hence even though he agrees that the dangers did not materialize, he is not able to admit to error, in retrospect, in any of the earlier public fears, or in any of his objections at each step in the relaxation of the guidelines. Not surprisingly, my review of his book is quite critical.

The conjectural dangers from recombinant DNA (rDNA) have failed to materialize, and the public's recent fear of this research has been replaced by a deep interest in its achievements and its promise. It is therefore time for a scholarly analysis of this remarkable affair, with its unprecedented degree of public involvement in a highly technical set of issues.

Sheldon Krimsky, a social scientist at Tufts University, has under-

---

Nature 301 (1983):543–544. A review of S. Krimsky, *Genetic Alchemy: The Social History of the Recombinant DNA Controversy* (Cambridge, Mass.: MIT Press, 1982).

taken the task of providing such an account, based not only on the published record but also on private discussions among scientists (recorded by others in oral histories). The book is scholarly in its chronological presentation and is heavily enough referenced to be a useful source. Unfortunately, however, it is dominated by the author's populist social perspectives: science is too dangerous to be left to the elitist scientists, and only direct public participation can protect the public interest. There is no sign of any reflection on the possibility that excessive public participation might have slowed the eventual resolution of the problem, or that the public's right to know should be balanced with a right to be spared from scaremongering.

The book is thus a curious mixture. The author tries hard to present the arguments on all sides. But although he avoids the strident tone of many earlier critics of rDNA research, he shares their suspicion of the scientific community. The resulting position is paradoxical. The present relaxation of the guidelines is not necessarily wrong, says Krimsky; but in describing each step on the way he offers an unsophisticated analysis that rejects the scientific evidence and judgments supporting the change, while treating with great respect even the most far-fetched contrary arguments. Hence this book is far from the judicious retrospective analysis that is needed. Let me suggest some of the points that such an analysis might consider.

First, one might ask whether or not the extensive public discussion of the hazards was a good thing. Many people would say yes. Scientists earned good marks for opening their doors to the public; all sides had their say, in the democratic tradition; and in the end reason prevailed. Nevertheless, I would agree with those who come out with a much less favorable balance-sheet. Large sums of money and much time were diverted from productive research; the United States Congress came close to enacting severely restrictive legislation, which would have been hard to reverse; the inroads into the traditional autonomy of science, and the imposition of an onerous bureaucracy (still present), set a dangerous precedent; the anxiety aroused was an unnecessary burden for the public; and the view of science as a threat was reinforced.

The crux of the matter is that there cannot be conclusive answers to questions about conjectural hazards. In the face of such uncertainty wide publicity is an invitation to emotional reactions at best, and to demogogy at worst. I would therefore criticize the format of the Asilomar conference, which was convened in 1975 to consider the hazards of rDNA research. A committee of 150, in the glare of world-wide publicity, is not an ideal instrument for evaluating technical issues. This conference, as described by some of those present, rather resembled a religious revival meeting. The outcome was a finely graded classification of risks, established as though it were based on sound science when in fact it was based on guesswork.

I have long wondered how such a distinguished group of scientists could have accepted such a metaphysical construction, culminating in a fear even of random human DNA as a dangerous material. (Krimsky fails to note that Joshua Lederberg and Jim Watson pleaded in vain against this course.) Krimsky's account provides a possible key. As he relates, the conference was built around the reports of working groups with research backgrounds in different classes of DNA. The members of the Animal Virus Group (chaired by Aaron Shatkin), accustomed to the risks of working with viruses, were the least worried. They handed in (with one dissent) a one-page report, recommending that research with viral recombinants in bacteria should follow already existing guidelines for work on the viruses themselves. In contrast, the report of the Plasmid Working Group (chaired by Richard Novick) filled thirty-five single-spaced pages. After expressing broad concern over environmental hazards and philosophical matters, it classified experiments into six levels of biohazard and corresponding levels of containment.

It is not clear how much the subsequent adoption of a detailed classification of risk, in the NIH guidelines, depended on this report. The Ashby Committee in Britain came out with a similar classification, but by a different mechanism. But whatever its ontogeny, the classification was in effect a certification of undemonstrated risks. It thus provided a foundation on which a handful of dissident scientists could arouse great public anxiety. What should have been a set of judgments about probabilities then degenerated into arguments about an inappropriate, potentially paralytic question: "Can you prove that the following could not happen?"

We must ask why the molecular biologists were willing to air publicly apprehensions that rested so heavily on guesswork and on the extrapolation of already uncertain knowledge. One reason was clearly an admirable sense of moral responsibility, coupled with political inexperience: they did not foresee how their conscientious descriptions of remote possibilities would eventually be interpreted as a conviction of imminent dangers. But perhaps the most interesting contribution of Krimsky's book is the recognition of an additional, cultural factor: the recent widespread loss of confidence in the authority of experts. Brought up to see authority in any form as elitist, and extrapolating from the real dangers from nuclear technology to the putative ones from biology, many young molecular biologists were ambivalent about the future social impact of their field. In this atmosphere the pioneers in recombinant DNA research must initially have felt quite virtuous in showing that scientists could now be open and antielitist. Moreover, the loss of nerve in the scientific community became widespread. Thus when the National Academy of Sciences decided to try to help it did not set up the traditional blue-ribbon committee. It held a public forum that gave equal time to all sides: the cautiously optimistic mainstream biologists, the handful of scientific Cassandras, and the political activists. This forum was not very helpful. Of course, it is not certain that a more traditional approach could have been more influential in such a charged atmosphere—but, in restrospect, one must wonder.

The molecular biologists also contributed to the problem by a lack of willingness to listen to biologists in other areas of research. Since molecular biologists had created whatever dangers might exist, and since they would be most affected by any restrictions, it is not surprising that it was they who assumed responsibility for assessing the risks, primarily by experimental tests. And, indeed, the favorable results of several risk experiments did contribute much support for the later relaxation of the guidelines. But from the start experts in other fields, closer to the problem of risk assessment, could have invoked principles that justified more reassuring judgments. Thus investigators of infectious disease deal constantly with pathogenic bacteria that are well adapted to survive in nature, and some were very doubtful that 0.1 percent foreign DNA in *E. coli* could create an even greater hazard. In addition, as Krimsky concedes (but without being convinced), two Darwinian arguments provided a theoretical framework for rejecting the early scary scenarios: introduction of foreign DNA will inevitably impair the genomic balance that is essential for survival and spread in nature; and since bacteria can take up DNA in nature the recombinants being made in the laboratory could not be a radically novel class after all. Unfortunately, it took time for all these principles to emerge and to be taken seriously, and meanwhile the course of the drama was already set at Asilomar.

Among the lessons that might be drawn I would suggest the following. First, the evaluation of risks must precede decisions about their acceptability; and while the public, through its representatives, should be heavily involved in the latter process, its involvement in the process of technical analysis is likely to be a hindrance rather than a help. Second, it takes time for scientists to see the implications of highly novel developments, and meanwhile the public does not benefit from exposure to transient, frightening hypotheses. Third, mass meetings of scientists are a much poorer mechanism for evaluating controversial issues than the traditional small committee. Fourth, nuclear technology, with its great economic and military pressure to underestimate the real dangers, is a poor model for assessing potential hazards in basic biological research. Finally, the search for absolute security is a will-o'-the wisp, diverting attention from real hazards and delaying real benefits. With highly conjectural hazards from scientific advance we can do no better than be guided by subjective probabilities, coupled with a sharp watch for early warning signs of tangible risk.

39

Inherent Limitations of Genetic Engineering in Man

As technical advances began to bring the prospect of gene therapy in man quite close, a concern that had lain beneath the surface of the earlier debate of recombinant DNA now emerged. At the request of leaders for this country's three largest religious groups President Carter established a presidential commission to consider the moral and legal implications of the novel powers that might emerge in this research area. This commission did a superb job of analyzing the issues responsibly. It recommended vigorous pursuit of the goal of somatic gene therapy (i.e., supplying the missing gene to individuals born with a monogenic hereditary disease), while at the same time warning against manipulation of genes in the germ line—though for reasons rather different from the ones that seem to me most cogent.

Representative (now Senator) Albert Gore, Jr., who was deeply interested in issues of biomedical research and ethics, proposed that Congress needed a permanent continuing commission to monitor further scientific developments in genetic engineering. I delivered the following statement at a hearing that he held on this subject. I was very much impressed by Mr. Gore's well-informed and thoughtful position, but nevertheless I questioned whether a commission with this narrow charge would be very useful. It would probably do no harm, but it seemed unlikely to accomplish much good. I suggested that concerns over hypothetical ethical issues arising from work with this particular set of techniques could be better handled within the framework of a commission with a broader mandate, covering the ethical aspects of all of biomedical research. It is gratifying that since then the senator has offered a bill to set up such a commission.

Many of the arguments in this paper had appeared thirteen years earlier, in a much more abstract atmosphere, in the first paper of this section.

House of Representatives Committee on Investigations and Oversight, Nov. 18, 1982.

The term "genetic engineering" is an unfortunate one, when applied to human beings. It carries overtones of a cold attitude toward people, as objects to be manipulated and remolded. Yet the goal of those working toward human applications of this technique is gene therapy—the replacement of the single defective genes that cause various hereditary diseases; and this aim is strictly within the humanitarian traditions of medicine. It is therefore essential, in discussing future prospects, to distinguish sharply between gene therapy and nonmedical uses of genetic manipulation. The nonmedical use that most people fear, of course, is the control of behavior for eugenic or political purposes.

Therapeutic and nontherapeutic applications not only differ in their aims: they also differ strikingly in the likelihood that we will have to deal with them in the foreseeable future, because they face very different technical problems. Unfortunately, however, most of the discussion of the subject has proceeded on the assumption that the two developments are indissolubly linked: if we developed the possibility of correcting the genetic cause of any disease we would also be creating the possibility of a Brave New World, with governments using the same techniques for deliberate interference with human nature.

This assumption became widely accepted when news of a spectacular scientific advance a dozen years ago, the isolation of a gene by Jonathan Beckwith and his colleagues, was accompanied by an even more newsworthy announcement: he regretted this success, because he believed that this line of research would soon lead to the power to manipulate human genes, and he did not trust our political system to ensure that this power would be used only to benefit the people. Quite apart from any preference for one or another political system, if I believed that we would indeed be reaching the capacity to use techniques of genetic modification to program human behavior in any general way I would also feel uneasy at that prospect, in the hands of any political system. But on purely technical grounds I disagree with the judgment that that power is in sight, or even likely as far ahead as we can look. My reasons are the following.

First, as some leading investigators in this field told you on the first day of the hearings, therapy even of single-gene defects is not yet around the corner, though replacement of defective cells is beginning to look feasible for those cells that function in widely distributed, loosely organized locations. These include the precursor cells in the bone marrow that give rise to the red cells and the white cells of the blood, and the precursor cells that give rise to our specific immune responses. But even here there are still many technical obstacles to overcome.

When cells are arranged in a highly organized way, as in the liver or the kidney, the prospect of replacing them, or of introducing a desired gene into them in a reliably controlled way, is much dimmer.

When we consider the technical problem of modifying behavior genetically we are dealing with an infinitely more complex pattern of cellular organization, involving a network of about ten trillion specific connections between about ten billion cells in the human brain. An enormous number of genes must be involved in the development of this circuitry, and any particular trait, such as intelligence or aggressiveness, must be influenced by a large number of these genes, interacting with each other and with the environment. It is therefore not surprising that we cannot yet identify a single specific behavioral gene, while we can identify several hundred that cause hereditary diseases.

Accordingly, the only prospect I can take seriously in this area, for the foreseeable future, would be a limited, vague alteration of behavior by influencing the level of various hormones. To achieve any more specific modification of behavior, involving altered circuitry, we would have to identify a set of genes that each have a small effect on a trait, isolate these genes, and transfer them together. Both the identification and the transfer would be very much more difficult than what we face with single-gene defects.

Another important difference is that behavior depends heavily on environmental influences as well as on genes. Accordingly, the effect of genetic changes on behavior would not be as sharply predictable as the effect of replacing an enzyme in a blood cell. An even greater obstacle arises from the difference in the time at which different genes act: most of the genes that contribute to individual differences in behavior must do so by guiding development of the intricate circuitry of the brain, and so they will have done their work before birth. And gene transfer could not conceivably rewire an already developed brain. In principle, one could circumvent this difficulty by replacing genes in germ cells. But this procedure would have little appeal, for one would be investing great effort to change some genes in a germ cell whose other genes were still an unknown, chance combination.

Finally, if some limited degree of genetic manipulation of behavior should ever become feasible, we must recognize that it would require cooperation of the subjects; and any population willing to cooperate in this way would already have lost its freedom. Moreover, this means of manipulating personalities would have to compete with other, less elaborate and less costly means, some already at hand. These include the familiar psychological methods, as well as possibilities provided by pharmacology, neurosurgery, and even eugenics (that is, selective breeding for the desired traits).

I cannot escape the conclusion that the rumors of the dangers of genetic blueprinting of behavior have been enormously exaggerated, and they have aroused much more public apprehension than the facts warrant. At the same time, it is clear that the development of effective gene therapy, even for a limited number of hereditary diseases, would be one of the greatest triumphs of medical science. And even if the procedure should prove to be expensive, its benefits would convert a miserable, helpless, and often brief life into a healthy one, and the costs would be amortized over that lifetime. This kind of research therefore deserves support and approval, rather than apprehension. It would be a tragedy if moral objections, based on fear of misuse of the same techniques, should interfere with such support and approval.

The fear of misuse that I have just discussed is rational, though based on an inaccurate perception of the facts. But some may also object to this research on the basis of a more abstract and less rational principle: that it is immoral or dangerous to "play God" and tamper with a person's genes, since these define his essential individuality. I find it difficult to take this objection seriously. It brings to mind a curious response when chemotherapy against syphilis was introduced by Paul Ehrlich early in this century: some objected on the grounds that the disease was God's natural punishment for illicit sexual behavior. That vindictive view did not prevail, and I am confident that any parallel view of hereditary defects, as inevitable acts of God, will not prevail either. On the contrary, as our power to identify and to correct defects increases our notion of rights will inevitably move to include the right to start life without a severe handicap, if it might be prevented.

If we agree, then, that gene therapy by itself does not present a moral problem, we still face the question of possible moral obstacles to the experimental introduction of these techniques in human beings. I would suggest that this problem is essentially the same as that faced by any new therapy, whether medical or surgical. There is always a tension between the desire to make a new mode of therapy available as soon as possible and the need to have its safety and efficacy thoroughly tested, first in animals. In resolving these problems the medical profession relies on a long tradition, now supplemented by the existence of bioethics review boards. I do not see any compelling reason for special legislative treatment for gene therapy. However, because this approach is so novel and has been so much in the public eye it should be handled by the profession with great care.

I would now like to consider another type of genetic manipulation of humans that has seemed much closer than gene replacement: cloning. This creation of genetic copies of an individual has been successfully accomplished with frog embryos, by implanting nuclei from their body cells into egg cells. Ten years ago it seemed self-evident that improvements in technique would sooner or later extend the procedure to mammals, and also to the copying of tested adults rather than of undefined embryos. This scientific advance, if possible, would be of obvious value in agriculture, in the copying of prize animals.

The extension of cloning to man would raise such serious moral problems that I would oppose it. However, it now seems doubtful that we will have to face the problem. For while we know that all the different kinds of cells in our bodies contain essentially the same set of genes, recent work strongly suggests that as embryonic cells give rise to fully differentiated cells some of their genes change. Adult cells may therefore never be able to initiate clones. If this proves to be true the cloning of mammalian adults may well be unachievable, for fundamental reasons rather than for reasons that might be overcome by advances in technique. Human cloning by nuclear transplant, aimed at copying individuals with already demonstrated traits, would then lose its potential interest—and its threat.

Embryos, on the other hand, encounter no such problem as a source of clones. In fact, such cloning has already been accomplished in mice, not by nuclear transplantation but by separating the cells of a very early embryo and using each to start a new embryo. But while this procedure is indeed cloning, in the technical definition of the term, it is cloning of an unknown new individual rather than copying of a known. It is thus not a violation of our natural process of reproduction, which makes each individual unique by randomly recombining genes from the two parents; it is simply amplification of the process of producing identical twins. The motivation for this form of cloning is not nearly as obvious as that for cloning adults. I therefore do not see a problem that merits legislative attention now, though one might conceivably arise in the future.

Finally, we should note that molecular genetics has already made concrete contributions to medicine in a third, rapidly expanding area: prenatal diagnosis of hereditary defects. This development is of great benefit to those parents who both carry a recessive defect in the same gene: instead of accepting the twenty-five percent risk of a defective offspring, or else denying themselves children, for several diseases they now have the choice of solving the problem by prevention, even though it cannot yet be solved by gene therapy.

Let me close by emphasizing the need to protect the search for basic knowledge from being restricted by those who fear possible undesirable applications. All knowledge is double-edged; and we simply cannot foresee all the applications, and all the social consequences, of any discovery. We can serve society best not by blocking any particular knowledge but by better controlling its applications. In the physical sciences we have begun to resist certain applications that are too dangerous to people or damaging to the environment. If such applications appear in biology they should also be prohibited. But in the application of molecular genetics to man, where enormously beneficial results are appearing, I do not yet see any threats from which society needs protection.

40

The Two Faces of Genetic Engineering in Man

This editorial summarizes, for a wider audience, the comments offered in the preceding testimony before a House of Representatives subcommittee. It also is a bit more explicit in its arguments against setting up a continuing commission specifically to monitor genetic engineering.

To those who deal with the victims of hereditary defects there can be no question that gene therapy—the use of genetic engineering to correct such defects—is an admirable goal, solidly within the traditions of medicine. Moreover, for the loosely organized cells of the bone marrow (though not for those of most organs) cure by implantation of genes in somatic cells now seems only a few years off. Unfortunately, however, the cold term "genetic engineering" has suggested to the public other, nonmedical potential uses of the techniques, such as reshaping our physiques or our personalities, cloning favored adults, or creating sub-human hybrids.

Two years ago the three main religious groups in this country sent President Carter a joint letter that viewed research in this area as a source more of danger than of benefit. The issue was referred to an excellent presidential commission, with Morris B. Abram as chairman and Alexander Capron as executive director. Its recent draft report, and subsequent congressional hearings under Representative Albert Gore,

---

Editorial in *Science* 219 (March 25, 1983). Copyright © 1983 by the AAAS.

Jr. (D-Tenn.), strongly supported the conclusion that gene therapy is a thoroughly legitimate goal. The problem has thus been handled in a much more sensible way than the emotional earlier debate over recombinant bacteria. Also encouraging is the restrained response of the major media to the recent announcement that the implantation of a gene for growth hormone into cells of mouse embryos had produced a gaint strain. Evidently gene therapy itself, separated from other kinds of genetic engineering, no longer seems to present moral problems different from those of other kinds of experimental therapy, and these are supervised by local bioethics committees.

On the other hand, both the commission and some participants in the hearings viewed changes in the germ plasm as more dangerous than somatic corrections because they tamper with evolution. But man has been tampering for a long time, both by domesticating and by extinguishing species. Moreover, as a form of preventive medicine, gene therapy in human embryos would have the same effect on the gene pool as an accepted approach: prenatal diagnosis, leading to selection for normal embryos in a family of carriers. The evolutionary argument thus does not carry much weight. However, there is a practical consideration that will deter responsible investigators from altering human embryos for a long time to come: the need for virtually perfect reliability. In somatic cell therapy a fifty percent cure rate would be a triumph, but manipulations of embryo cells that damaged even one child in a thousand would be intolerable.

Although the commission did not consider the conceivable nonmedical uses an immediate threat, it recommended the establishment of a body to watch future advances and protect against their misuse. But some interventions, as we have seen, are too dangerous to apply to humans, while others are distant or impossible. In particular, the possibilities for genetic control of behavior, as in Aldous Huxley's *Brave New World*, seem much more limited than those for the cure of monogenic diseases, both because behavioral traits are polygenic and because most genetically determined differences between individuals are laid down in the brain circuitry before birth.

It thus appears that a special continuing commission on genetic engineering might find itself watching only for developments that either are very distant or are too dangerous to try. If so, it would have little to do, and it might then be tempted to become a busybody, imposing federal restrictions on activities that are better regulated on the local scene. On the other hand, the existence of some mechanism for continuing surveillance of genetic engineering could have real value in protecting the public from unwarranted anxiety. Perhaps the best way to achieve this end, while avoiding undue interference, would be to assign the task not to a special body but to one with wider responsibilities for biomedical ethics.

41

Genes and Souls

In contrast to the excellent report of the presidential commission on Splicing Life, *the indefatigable Mr. Jeremy Rifkin muddied the waters by managing to get over fifty clergymen, many very prominent, to sign a resolution demanding a legislative ban on certain kinds of genetic intervention. Though the impressive list of signers gained wide attention for this statement, the clergymen soon learned that they had supped with strange company, for Rifkin accompanied the release of the resolution to the press by a long piece, written in his usual apocalyptic style, which he labeled a "theological letter" (though it had no evidence of either input or approval from any theologians).*

I find the arguments that have been offered against germline intervention in humans unconvincing. However, in this OpEd article I replace these by what I believe are much stronger arguments for the same conclusion. First, the same goal can be reached by much simpler means. Second, in germline intervention the danger to the future person is so great that no responsible medical investigator would be interested in carrying out this procedure, or would be able to obtain approval, for at least as far ahead as it is profitable for legislative bodies to look.

While genetic engineering in humans seems close enough to justify public concern, we must not be swept away by fear of exaggerated dangers.

Unfortunately, a broad spectrum of clergymen recently demanded a total ban on attempts at one kind of genetic engineering, germline in-

---

OpEd *New York Times*, June 28, 1983.

tervention—that is, insertion of genes into a cell that will become an embryo. It was argued that a distinction cannot be drawn between medical and eugenic uses, and that we should not try to eliminate defective genes because we have no right to tamper with evolution.

In contrast to the apocalyptic tone of this discussion, a presidential commission has realistically analyzed many conceivable dangers. Its excellent report, issued in December, strongly supported the goal of somatic gene therapy—that is, insertion of genes into body cells but not into germ cells (sperm or eggs). But the commission also expressed deep concern about changes perpetuated in future generations, because such intervention would open up the awesome prospect of directing future evolution of the species. The commission did not call for a ban, but its position may have encouraged such a call. It is important to examine this evolutionary argument carefully and to identify the real issues.

Let us consider parents who both carry a recessive gene for sickle cell disease, along with the corresponding normal gene. A child who inherits the defective gene from each parent will have the disease, but a child with a single defective copy will not. Three methods for preventing or curing the disease are conceivable: identifying the double defect in an embryo by prenatal diagnosis, thereby giving the parents the option of abortion and another pregnancy (prenatal selection); replacing the defective gene in somatic cells after birth; or replacing it in the embryo.

It is easy to see that all three approaches would influence evolution. Prenatal selection would encourage parents to produce carriers, while somatic correction would produce people with a healthy body but a double defect in their germ cells: Both would increase the frequency of the sickle cell gene in the next generation. Germline correction, in contrast, would decrease the frequency. If there were no other considerations, what sensible person would not prefer germline intervention, provided it is limited to therapy?

But there are other considerations. First, the real long-term danger is that genetic engineering might be used not only for therapy but also to "improve" or blueprint people, according to somebody's plan. But somatic cells might also be manipulated for this purpose. And while the range of conceivable effects is broader for germline intervention, the important line to draw is that between medical and eugenic uses, rather than between somatic and germline cells.

Fortunately, we are unlikely to face eugenic uses in this century, because the traits one might be tempted to manipulate, such as memory, intelligence or motor skills, are so complex, and involve so many genes, that the prospects for their meaningful control are very distant.

While the evolutionary and the blueprinting arguments thus prove to be weak, there are other, overwhelming reasons not to proceed with germline intervention. First, where a corrective gene enters a chromosome it may interrupt some other important gene, producing a new hereditary defect. Moreover, while somatic therapy, in a person already sick, warrants a substantial risk, the manipulation of a cell that is destined to become an infant would require very stringent standards of reliability, far beyond what is in sight today. Finally, for a few diseases, so far, we already have a safe and simpler alternative method for reaching the same goal of preventing defective births: prenatal diagnosis and selection. For all these reasons, the motivation for altering the genes in an embryo is very slight.

If we do not need a ban on germline intervention, do we need a permanent commission to monitor human genetic engineering, as Congress is considering? Probably not. A commission with such a narrow assignment might have to invent things to do. Its existence would likely arouse false fears, for there are always people eager to stir up anxiety, and genetics is a favorite target. Scary scenarios about manipulating our inner nature—our selves or souls—have wide appeal, and the resulting pressures could interfere with beneficial medical research.

We should surely continue philosophic discussion of human applications, on the excellent base provided by the presidential commission. But since we are dealing with potential treatment of individuals, and not with possible large-scale effects on the gene pool or with epidemics, we can afford to postpone legislation until a concrete problem comes into view.

42

Science, Fanaticism, and the Law

After the prolonged debate of the 1970s over the hypothetical danger from recombinant bacteria had subsided, and the National Institutes of Health guidelines were gradually relaxed, it seemed that public anxiety over genetic engineering had been laid to rest. However, when studies on agricultural applications of recombinant bacteria or plants recently reached the stage of requiring field tests a new wave of objections arose. These were based, like the earlier wave, on hypothetical scenarios of very low probability. However, this second round did not arouse a strong public response, and the Recombinant DNA Advisory Committee of the NIH promptly approved the first proposed field test of a recombinant bacterial strain. Nevertheless, Mr. Jeremy Rifkin, the professional opponent of genetic engineering, obtained an injunction against this release to the environment.

Rifkin succeeded because he managed to obtain the support of a few respected ecologists, thus creating for the courts the impression of wide division of opinion in the scientific community. However, one of these ecologists later published a retraction, conceding that the organism in question (which had been genetically deprived of a virulence factor) could not reasonably be considered dangerous. What concerned him was the possibility of more threatening future developments in this area, and the conviction that his profession should be more heavily involved in the evaluation process. This kind of support for a legal action seems to me odd, for it is my understanding that courts are expected to judge a case strictly in terms of its specific features, and not in terms of possible future related cases.

The main concern of the ecologists is that the novel recombinants might spread in an uncontrollable manner, like those naturally occurring or-

Genetic Engineering News (July/August 1984):4.

ganisms that have become pests when transplanted to a new continent. Here I discuss why I believe these ecological misfortunes, involving organisms that have already been selected in nature for their adaptation to the natural environment, are not close models for the behavior of an organism that has been jerry-built in the laboratory and is selected by man for growth under conditions of cultivation. Moreover, many earlier variants, created by classical methods of plant breeding or bacterial strain improvement, have been tested and licensed for commercial distribution, and they have not caused any ecological damage. It is not clear why modifications created by the recombinant technique need be treated any differently.

Of course, it is understandable that ecologists should be pressing for more extensive involvement of their profession in evaluating the release of any engineered organisms. Unfortunately, they have not been able to come up with convincing evidence that the danger is significant, or that they have a concrete program for providing the firm predictions and estimates that the law is asking for. Nevertheless, the current dialogue between ecologists and molecular biologists will no doubt be educational for both. Moreover, if we are fortunate the resulting legal compromise will profit from the experience of the 1970s, and we may be able to avoid the wasteful repetition of a cycle of excessively stringent regulations followed by relaxation.

This paper comments briefly on the scientific issues in the specific case that Rifkin attacked. It also quotes some statements in his most recent book that may provide insight into his aims.

In granting an injunction against a proposed release of recombinant bacteria, Judge Sirica's surprising decision has blocked, on extraordinarily weak grounds, a legitimate, responsibly evaluated scientific experiment. He has thereby set a dangerous precedent, and he has also given Jeremy Rifkin's pseudoscientific, apocalyptic predictions more credibility than they would otherwise have. It is therefore important for the scientific community to respond.

Why the Suit Is Frivolous

The law requires an environmental impact statement only for "actions *significantly* affecting the quality of the human environment"; and the judge found "several areas of *plausible* environmental concern." In fact, it would be hard to find a less plausible case than the Berkeley ice nucleation trials.

First and most important, mutants of *Pseudomonas syringae* that no longer promote ice formation are not new to the environment. They occur naturally but are rare, because they cannot survive as well as the parental strain. If an environmental niche does exist for such mutants they have long since found it, in the eons during which they have continued to arise. Moreover, similar mutants, obtained in the laboratory after simple mutagenesis, have already been tested in the field, without harm. To produce a better defined and more stable mutation the current experiment uses recombinant DNA, and because of this small technical modification the whole experiment now required approval of the NIH Recombinant DNA Advisory Committee (RAC).

There is thus no reasonable scientific basis for the claim, in the suit, that the altered bacteria might spread. There is even less basis for the fanciful predictions of dire consequences if they should spread—for example, interference with cloud formation, or harm to those plants that are naturally frost-resistant. It is not surprising that the NIH committee, containing outstanding scientists, found no significant danger in the experiment.

It has also been argued that the release of modified bacteria might create pests, like starlings or the gypsy moth. However, this analogy is irrelevant. Such explosions have occurred only when a species was transferred to a new continent, where it no longer encountered the ecological restraints that held it in check in its native habitat. But a bacterium that is modified genetically and released into its original environment, as in the ice experiment, will not encounter such an ecological vacuum. Moreover, unlike higher organisms, bacteria can be wafted in the air, and identical species are found on all continents (except for species adapted to a unique ecological niche).

Rifkin's Aims

Those who are impressed by Rifkin's approach, including Congressman Albert Gore, Jr. (whose enthusiasm is displayed on the dustjacket of Rifkin's recent book *Algeny*), should look more closely into his aims. When his group of activists forcibly took over a National Academy of Sciences Forum on Recombinant DNA, in 1977, his ideology seemed to be simply that of the anti-establishment counterculture. Now, however, it is clear that he is motivated by a much more personal mystique, illustrated by the following quotations from *Algeny*.

On Darwin: "Perfect efficiency would amount to having everything at one's disposal that could possibly be produced without having to exert any energy whatsoever . . . [According to Darwin] evolution was always advancing toward the perfectly efficient organism, meaning the perfectly self-contained organism, meaning an organism invulnerable to all outside influences, meaning an organism remarkably similar in constitution to God." Finally, "there is no doubt that [the] attacks . . . eventually will triumph, leaving Darwin a lifeless corpse."

And in conclusion: "There could be no lonelier place than a biologically engineered world. That's why even if only one living creature were left unscathed in a world brimming over with biological facsimiles, we would reach out to it, embrace it, touch it, marvel at it, with a peak of emotion that all the replicas together could not hope to tap in us."

One would not expect this kind of rhetoric to have any influence on the course of science. Even when toned down, in a legal strategy, Rifkin's fantasies can hardly long delay useful applications of genetic engineering. Nevertheless, his present attack, supported by an antiscience public mood and by political interests, could lead to a replay of the cycle of anxiety, sterile debate, bureaucratic regulation, and eventual recovery that we lived through in the late 1970s.

It is therefore important to recognize that NIH's RAC has developed, through the past half dozen years, a sober, realistic appraisal of the hypothetical dangers that earlier loomed so large, and it has done very well in progressively relaxing the guidelines. The resulting benefits, and the complete lack of harm, speak for themselves.

Procedures and Criteria

Rifkin, Judge Sirica, and some editorial writers delude themselves in thinking that an environmental impact statement would solve the problem by providing clear, general, uniform standards. The experience of RAC shows that in this fast-moving field there is no substantial basis for such firm standards, any more than there was for the elaborate P1 to P4, EK1 and EK2 scheme of the initial NIH guidelines.

The whole problem is likely to resolve itself into the political question of whether approval of release of recombinant (or otherwise modified) microbes should continue to be in the hands of RAC (perhaps expanded to include a soil microbiologist and an ecologist), or whether a new agency with different expertise is needed.

I submit that if the judgments of the safety or danger of each organism really depended primarily on data on the kinetics of its survival in various environments, it might well be logical to rely on one set of experts to judge potential medical dangers and on another set for environmental dangers. But in fact, it is impossible to test all the ecological niches into which an organism might spread, and it is clearly unrealistic to expect to test extensively each of innumerable recombinants that are being made.

Hence, while it is reassuring to demonstrate that some novel organisms disappear rapidly in some experimental settings, in the last analysis the judgments are made, and must be made, largely on the basis of fundamental Darwinian principles. These are the same for the spread of any bacteria—in the human population, on plants, or in the soil. The key principle is that the organisms found in nature have been selected over millions of years, from a virtually limitless supply of variation, for their adaptation to some ecological niche; and any genetic modification introduced in the laboratory is infinitely more likely to impair than to improve the adaptation, unless the environment is also changed. (For example, widespread use of antibiotics selects for resistant strains.)

To be sure, as in the medical area, there could be dangers, at least of local harm, in experiments that introduced virulence factors or altered the host range of a pathogen. But no one is proposing to release plant pathogens. (Biological warfare research, of course, could be the exception—but here neither EPA nor NIH would have control.) If release of pathogens should be proposed, that would be the time for an environmental impact statement. Meanwhile, the case-by-case judgments of RAC do not deserve the skepticism that has recently been stirred up.

Though Judge Sirica emphasized that he was not evaluating the scientific arguments, he did judge that there were plausible causes for concern. With that precedent any variants of the ubiquitous useful bacteria may be seen as enemies, requiring elaborate public exoneration whenever demanded by fanatics. We would then be following a script out of the pages of Lewis Carroll—or Jonathan Swift.

43

On Gould on Rifkin

In an excellent article on which this brief letter comments, Stephen Jay Gould dissected the fallacious arguments of Rifkin in much greater detail than I had done, and with greater finesse. At the same time, he surprised me by agreeing with Rifkin's basic view that genetic engineering threatens the integrity of the natural living world. The assumptions underlying this view seem to me incorrect, and since the issue is a fundamental one for the public's view of genetic engineering it seemed worthwhile to set forth my reasons.

Stephen Jay Gould has done a fine public service, and a masterly job, in dissecting the pseudoscience and the pretentious philosophy in Jeremy Rifkin's *Algeny* ["On the Origin of Specious Critics," Jan.]. Moreover, while Gould's main aim is to defend evolutionary biology against distortions in that book, he also explicitly defends the use of genetic engineering for such valuable purposes as crop improvement. Yet at the same time he expresses sympathy for Rifkin's basic view: that genetic engineering threatens the integrity of the natural living world. Unfortunately, this ambiguous position may seem to some readers to offer support for Rifkin's appeals to the public and to the legal system. These activities are much more influential than Rifkin's "philosophical" writings. I must therefore question some assumptions that seem to underlie Gould's position.

First, Gould engages in a bit of hyperbole in speaking of the power of "altering life's fundamental geometry and permitting one species to

Discover (April 1985):85.

design new creatures at will"—for it cannot in fact be done at will. As he knows very well, in the evolution of a new species the new genes have to fit into the total pattern, so that all the parts of the organism can develop and function in a coherent way, just like the parts of a machine. Otherwise the organism will be less fit and may not even live. This requirement of coherence inevitably restricts the range of useful variation that plant and animal breeders will be able to produce by molecular manipulation of genes. The possibilities for modification will be expanded, but will not be unlimited.

On a more fundamental point, Gould states that our power to manipulate DNA raises a "deep and distant issue." This is clearly true, especially of applications to human beings. But I am not sure what he means when he says, "I do not disagree with Rifkin's basic plea for respecting the integrity of evolutionary lineages." Mankind has already violated that integrity by developing hybrids, such as the mule and the tangelo, and these useful developments have hardly been evolutionary catastrophes. Moreover, such hybrids create much more extensive genetic recombination than we can expect from molecular techniques, which can introduce only one gene, or a few, into the hundreds of thousands present in an animal or plant. Hence it is not clear why these new techniques would threaten evolutionary lineages. To be sure, future developments in cell fusion may make it possible to extend hybridization to a wider range of species—but the history of horticulture encourages us to expect this extension to be used to add to the rich diversity provided by evolution, rather than to threaten it.

It seems clear that Gould's real concern—which I share—is not the addition of novel organisms. It is the rapid recent increase in the extinction of existing ones. But the causes have been the expansion of the human population and the spread of technology—not successful Darwinian competition of domesticated organisms, obtained by artificial selection, replacing wild organisms produced by natural selection. Why the new kinds of domestication, by molecular techniques, should be a greater threat is not clear. It is not the nature of the selected organisms, but how we use them, that will count.

44

Profit Sharing between Professors and the University

Lying somewhat outside the range of topics of the rest of this volume, this editorial is concerned with problems that have been generated by the commercial promise of genetic engineering, rather than by its threats. The issue became prominent when the administration at Harvard University proposed to set up a biotechnology company to build on a discovery by a faculty member, Mark Ptashne. The proposal went smoothly through earlier committees, but when it was presented to the whole faculty, in an open-ended form that permitted unlimited scenarios, the objections of a few members soon grew into a storm in the press. President Bok prudently withdrew the plan.

Meanwhile many other academic institutions have made potentially profitable arrangements of various kinds, with no obvious harm. Nevertheless, I now agree that the proposed intimate involvement of Harvard in such a speculative enterprise would not have been wise. But I find it sad that the violent reaction to the prospect of this kind of involvement precluded the possibility of retreat to a more modest one.

The main theme of my paper is that when faculty members set up profitable enterprises based on their professional activities they are benefitting from tangible and intangible contributions of their university, and so it seems fair that some share of the profits should go to the nurturing institution. One possible device is a gift of stock, but without managerial responsibility. In an even more straightforward mechanism, one newly formed company has allocated a fixed percentage of its future profits to its board of scientific advisers, for distribution to their universities. The justification for this arrangement was simply that the company benefitted great-

New England Journal of Medicine 304 (1981):1232 .

ly from its location near a large university community. If a company were originated by faculty from a single university it could easily make a similar arrangement for future benefits directly to that institution.

The intellectual yield from our country's investment in fundamental biologic research has been tremendous. Through the fusion of genetics with biochemistry, and through the development of ingenious techniques for exploring the range of dimensions that were formerly hidden between microscopy and chemistry, we have acquired a remarkably coherent picture of the universal features of cell structure and function. Moreover, Darwin's theory can now be considered Darwin's law: We understand why information cannot flow back from phenotype to genotype, and we can measure evolutionary distance directly in terms of DNA sequences. Yet, however gratifying these triumphs of human intelligence and imagination may be, interest in science as a cultural enterprise was not the main reason for the generous flow of public funds. Valuable applications were promised and were expected.

Nevertheless, unlike earlier experience with the biochemistry of small molecules (amino acids, vitamins, hormones, and antibiotics), the practical payoffs from molecular genetics were disappointingly slow: the intracellular macromolecules of the molecular biologist could not be translated easily into medical prescriptions. The resulting impatience of some legislators was understandable, but the picture has now changed dramatically. Emerging from the integration of three decades of fundamental and even esoteric research, the recombinant-DNA technology now promises innumerable applications in agriculture and in energy production as well as in medicine. Molecular biology has thus burst into the age of high technology, and the resulting acute speculative fever has perhaps been exacerbated by the earlier celebrity of recombinant DNA as a presumed menace.

The new commercial possibilities present universities with both opportunities and risks. This situation has been brought into sharp focus by a recent proposal of the Harvard administration, which is discussed in Barbara Culliton's article in this issue of the *Journal*. In my analysis of these problems, I shall proceed on the following assumptions: first of all, that the university community has a social obligation to try to promote technology transfer, but that it should do so in ways that will not jeopardize the search for knowledge for its own sake; secondly, that in this country the options will continue to lie within the framework of a system of private enterprise; and thirdly, that because of the changing pattern of government support of scientific research, universities are obligated to try to find additional sources of funds, both to preserve their autonomy and to prevent a waste of trained scientific talent. All these assumptions may be questioned, of course, but such questions are beyond the scope of this discussion.

Patents

Patents have long been used to provide income for universities. For example, the royalties of the Wisconsin Alumni Research Foundation since the 1920s have made biochemistry an unusually strong field at that university. In general, however, biologists have been much less interested in patents than chemists or engineers have—perhaps because their work has only rarely lent itself to applications. Moreover, in medicine the earlier tradition of charity to the poor raised additional barriers to commercialization. Accordingly, for many years the Harvard Corporation required that any health-related discoveries in its laboratories be dedicated to the public. Meanwhile, there have been major changes in the economics of medical care and in the attitude of federal agencies toward patents. In addition, it has become clear that the absence of a patent often impairs, rather than promotes, the availability of a useful product to the public. Accordingly, in 1975 Harvard decided that it would take out patents and would transfer a modest fraction of any royalties to the inventor.

Some object to even this degree of university involvement, on the grounds that it may lead to preferential treatment of certain faculty members. However, this problem is not unique to patents. Grants bring in a large overhead. Moreover, faculty salaries are often derived from grants to individual faculty members, and the influence of external funding on appointments may then be considerable. In contrast, any royalties from patents can be distributed without externally imposed restrictions. It is thus hard to see why patents are likely to bias appointments more than grants and contracts do.

Another criticism of patenting is that it encourages secrecy. However, as is well known, secrecy is widespread in highly competitive fields of even the purest research. To be sure, patentability may provide an additional incentive to secrecy over the short term, but in the long run patents eliminate the need for secrecy, since after the date of filing of a patent application, like the date of submission of a publication, the information is released for free discussion and for noncommercial use by others. Indeed, in industry unpatentable information is the main body of trade secrets.

The history of antibiotics provides an interesting lesson. When Ernst Chain isolated penicillin he urged that it be patented for Oxford University, but the British establishment in academic medicine refused. As a result, Oxford received nothing; the British were soon paying royalties to American firms. In contrast, a few years later the royalties from Waksman's discovery of streptomycin at Rutgers, and from Umezawa's discovery of kanamycin in Tokyo, were used to found and support excellent research institutes.

Private Corporations and the Special Problems of Biology

Another well-established mechanism for the commercial exploitation of science—potentially more profitable for the professor—is the formation of a private corporation. That a number of molecular biologists have initiated such undertakings is not surprising, especially since research in this field has been characterized by unusual boldness in moving to challenging new problems.

Universities have treated such activities much as they treat industrial consulting—a practice that they often encourage because, like part-time practice in medicine, it helps to retain valuable faculty. Consulting time is often limited, either by a formal rule or by an informal agreement, to one day a week. However, the development of a new company is likely to require much more time. Moreover, the impact of such an involvement on academic activities cannot be measured entirely in terms of formal hours of work; a change in what the professor is thinking about when showering or driving may have an even greater effect.

An excessive diversion of time may be only a temporary stage at the start of a new business. Accordingly, in this area universities are justified in continuing their tradition of flexibility and patience in supervising the daily distribution of faculty time. On the other hand, the problem of a substantial diversion for a long period under the umbrella of a full-time university salary cannot be ignored, and I shall return to it.

The collegial academic atmosphere is likely to be even more seriously harmed by competition between faculty colleagues who are associated with different companies. The tradition of industrial ties in chemistry and in solid-state physics provides a somewhat reassuring model, but there are important differences, not only in the traditions but also in the content of the fields. Specifically, in recombinant-DNA research the competition may well be much more intense than it is in these other fields, at least for the present, because the range of problems is so much narrower: many groups are inevitably seeking the same product, such as insulin or interferon.

Two other differences between biology and chemistry are also pertinent. In the first place, biology often asks more philosophic questions that are remote from potential applications. Secondly, because of the extraordinary complexity of biologic material, major breakthroughs frequently depend on a happy accident, an unexpected observation, or an unpredictable implication of distant findings. If the search for wealth diverts too many of the best biologists in the next generation from undirected exploration of the nature of life, something precious will be lost.

Clearly, these are serious problems. If research in this new industry becomes more autonomous and developmental, perhaps like research in antibiotics over the past decades, its present resemblance to university research will dwindle, and so these problems may be only temporary. However, antibiotics arose from an accidental discovery, and their pursuit has remained largely empirical, whereas recombinant DNA has extensive theoretical roots. Hence, its applications are likely to continue to depend heavily on advances in academic laboratories.

The Harvard Proposal and the Future of Institutional Profit Sharing

A major difference between the two mechanisms of technology exploitation described above is that universities have generally shared in profits from patents on discoveries made in their laboratories, but not in those from companies stemming from such discoveries. However, some European laboratories have recently extended profit sharing to the second arrangement. A similar proposal was made at Harvard last October but was soon withdrawn. It elicited a strongly unfavorable response from some of the faculty and from editorial writers in the news media. Now that the tempest has subsided, it may be useful to reexamine the issues.

The proposal arose when Harvard was considering the possibility of patenting a discovery of Professor Mark Ptashne in recombinant DNA. The Harvard administration suggested the alternative of setting up a company with outside venture capital and with the university given a minority share (which was subsequently said to be ten percent). The company, in return, would have the rights to any patents on Ptashne's discoveries held by the university. In the memorandum that opened the discussion the Harvard administration carefully spelled out a number of pitfalls in this kind of venture, and it asked the faculty to consider the abstract policy issues and principles involved. However, it did not specify any details of the proposed arrangement or even mention Ptashne's name; it simply informed the faculty that it would be making a decision on a specific arrangement within three weeks.

With the benefit of hindsight, one can see several aspects of the presentation that promoted resistance. The linkage to a concrete decision foreclosed the leisurely philosophical discussion that was requested, since it gave a sense of urgency to those who were opposed. Moreover, the lack of details about the proposal aroused mistrust, and vagueness about the role of the university in the proposed company gave rise to the widespread misapprehension (especially in the news media) that Harvard was actually going to operate it. The memorandum emphasized the advantages of having the university protect the interests of a faculty member; but willing acceptance by the faculty would have required a degree of confidence in institutional authority that is not universal today. Concern was also heightened by the earlier, wild public response to the Genentech stock offering, which had capitalized that company—as yet without a salable product—at over $500 million. Many faculty members gagged at the prospect of having the university linked with a similar caricature of the capitalist system.

In addition, many faculty members resent the venality and secrecy that has arisen in some laboratories performing recombinant-DNA research. This resentment clearly intensified the reaction, and it led to the understandable conviction that the university should not appear to condone and perhaps even encourage this pernicious development. However, the hostility may have been misplaced. The problem arises from the lucrative potential of the work, not from the possibility of university ownership of shares.

In the end, two dangers seem to have caused the greatest concern: pressures on faculty in their choice of research, and favoritism of the university toward financially productive faculty members. The memorandum unfortunately presented both these problems as though they were novel, instead of comparing them with the similar problems associated with other sources of funds. As a result, the issues were analyzed rather unrealistically, and an idealized conception of the university was defended: the institution was presumed to be entirely free of restrictions on how it distributes its income in supporting the preservation, advancement, and dissemination of pure knowledge. In fact, research grants from government, foundations, and industry—and many endowments in support of professional chairs or specialized institutes—do not provide such freedom.

With respect to the basic issues of favoritism and freedom of research, it is instructive to compare the rejected proposal with the recent twelve-year grant of $23 million from Monsanto Chemical Company to Harvard Medical School, as described in Culliton's article. This grant has expanded the facilities that are available to the recipient professors, outside the academic control of any department, and it surely commits the recipients to a given line of research. In contrast, dividends from equity in a company need not have either of these consequences, and such funds could be distributed by the university much more freely. Of course, one could consider the Monsanto grant a poor model to follow. However, it seems unlikely that universities will find better terms in their present search for industrial support. The area of research is appropriate and remains basic, freedom of publication is unencumbered, and the institution gains some permanent resources. The company selects the investigators and their area of work, and it has favored access to the results before publication and to patents; but it seems unrealistic to expect large-scale industrial support without such an exchange. In addition, to the extent that dwindling federal funds are replaced in this way, the research community as a whole benefits.

Universities have had a long history of negotiations, especially in medical schools, over academic activities and positions linked to private gain. The members of the Harvard Faculty of Arts and Sciences demonstrated little awareness of this history in their recent discussion. The emergence of professors as entrepreneurs raises a wide range of problems, but the only one discussed was the prospect of having the university involved as well. In particular, no questions were raised about existing companies that do not contaminate (or benefit) the university—for example, one in biotechnology that was recently initiated by Professor Walter Gilbert and one in economic consultation that was founded by Professor Otto Eckstein and was recently sold for about $100 million.

In the absence of limiting ground rules, it was perfectly proper for enterprising faculty members to have set up such unshared private corporations, but I suggest that the rapid expansion of such activities now demands a broader look. Medical schools have long faced a similar problem with full-time salaried faculty members who collect fees from private patients. Many solutions have been tried, ranging from complete transfer of the money to the institution to no transfer. Unfortunately, this experience does not offer any ideal, universally accepted model for other faculties beginning to face a similar dilemma. Nevertheless, it has certainly not been obvious that complete retention of the income by the faculty member best serves the university, or that it represents the fairest possible arrangement.

Conclusions

We may all regret the loss of the more Arcadian atmosphere of the past. However, if universities are to protect their financial base in order to advance their academic goals, nostalgia will be no substitute for imaginative adaptations and a tough-minded attitude. There are surely risks in developing industrial connections, but they must be balanced against the increasing financial insecurity of universities today, and against a monolithic dependence on an often unsympathetic government.

There is also a question of simple justice. The facilities, the atmosphere, and the financial support of universities have provided an essential background for many commercial developments, and the continued connection of the entrepreneur with the university, like the connection of a physician with a teaching hospital, often gives him a good deal of prestige. It therefore seems just that the university, which can no longer afford to be in the position of a generous parent, should in return receive a share of the profits. In addition, such profit sharing would respond to a widespread and cogent criticism of the present system: that it unfairly allows professors to become rich through developments stemming from tax-supported research. There is still appeal in the basic concept of dedicating a medical discovery (especially a tax-supported discovery) to the public interest, and distribution of part of the profit to the university surely serves the public interest more directly than does distribution only to the other participants.

One could argue that licensing patents is a less entangling way to reimburse the university than is the sharing of equity. However, equity in a corporation offers not only the possibility of a larger income to the university; it may be even more important as a means of ensuring continued benefits from future discoveries. Once a professor had begun to direct research in an industry as well as in an academic laboratory, he or she would no doubt be tempted to shift to the latter any brand of the academic work that appeared potentially patentable, thus foreclosing any future possibility of royalties for the university.

Finally, profit sharing could have certain mutually advantageous by-products that were not mentioned in connection with the Harvard proposal. For example, if the industrial laboratory was nearby, which would be convenient for all concerned, a financial interest by the university could eliminate the question of recompense for access of company scientists to libraries and seminars. Similarly, the specialized instruments and facilities for large-scale preparations in an industrial laboratory could occasionally be useful for university researchers.

I have suggested that various arguments against the proposal of the Harvard administration were not convincing. Nevertheless, the possibility of conflict of interest is real, as is the problem of keeping the business connection at arm's length from the academic activities of the university. In addition, there remains a serious moral issue. Given the rules of the game, the scientist-entrepreneur is free, within the restrictions of the Securities and Exchange Commission, to convert paper profits into a fortune by selling stock at an inflated, speculative price. Similarly, it is legitimate for a university's investment managers to seek capital gains in the open market from fluctuations in the price of such securities. However, if its connection with a company increases public confidence, a university has an additional responsibility not only to protect its reputation but also to protect the public against the creation of a financial bubble.

If the arguments for profit sharing prevail, universities may have to establish requirements for some such arrangement before allowing a faculty member to hold an important position in a corporation. Indeed, even consulting is not necessarily sacred, any more than are fees from patients: investigators at the National Institutes of Health are not allowed to retain consulting fees or lecture honorariums, and President Hutchins once introduced such regulations at the University of Chicago. Of course, this rule effectively discourages consulting, and hence technology transfer; but an arrangement for sharing might not. As the commercial applications of biology grow, there will be room for imaginative experiments, perhaps with buffering organizations like the Wisconsin Alumni Research Foundation between the university and the corporation.

Whatever the main reason for the negative reaction to the Harvard proposal—whether it was the manner of its presentation or the devotion of some faculty members to an idealized conception of alma mater—this reaction is clearly not the last word on the subject. A new company in France, Transgène, has distributed equity to the Pasteur Institute in Paris and to the University of Strasbourg, and in England a national biotechnology corporation will be sharing profits (and results of research) with the Medical Research Council Molecular Biology Laboratory at Cambridge. In this country, several universities and research institutes seem to be moving rapidly in the same direction. For better or worse, the objections to such arrangements may fade even more rapidly than did the earlier objections to patenting.